"When my radioman told me that Lieutenant-General Moshe Dayan was with another Marine unit on a nearby hill in the summer of 1966, I didn't believe him at first. But, indeed, Dayan was there. He came to Vietnam to study modern warfare in the field, where and when it was happening. His astute and forthright observations, translated into English for the first time, should be read by any person interested in the Vietnam War and the American experience in Vietnam."

Bing West
Marine combat veteran, Vietnam
Former Assistant Secretary of Defense
Author of *Small Unit Action in Vietnam, Summer 1966; The Village, The Strongest Tribe, The Wrong War, The Last Platoon, Call Sign Chaos* (with Jim Mattis), and others

MOSHE DAYAN

Foreword by GENERAL CHARLES C. KRULAK, USMC (RET.)

VIETNAM DIARY

Translated and Edited by

PETER C. APPELBAUM

STONE TOWER PRESS

Original Hebrew edition,: Yoman Vietnam, (*transliteration*). Copyright by the Dvir Co. Ltd., Tel Aviv, 1977.

Permission for English translation granted by Rachel Dayan.

Cover image: Photographer: Israel Press and Photo Agency (I.P.P.A.) / Dan Hadani Collection, National Library of Israel / CC BY 4.0. Image used with permission. (990044326700205171).jpg

All other images also Dan Hadani Collection, National Library of Israel. Images used with permission.

Stone Tower Press
7 Ellen Rd.
Middletown, RI 02842
stonetowerpress.com

ISBN: 979-8-9940926-6-8

Formatting and cover design by Amy Cole and Emily Olson, JPL Design Solutions

Printed in the United States of America

This translation is dedicated to my father Hymie Appelbaum z'l and my wife's father Murray Rosen, z'l. Both served during World War II, my father as physician in the South African Army Medical Corps in Egypt and Italy, my father-in-law as an artilleryman under General Patton during the Battle of the Bulge

TABLE OF CONTENTS

Gen. Charles C. Krulak, then a young officer, escorting Moshe Dayan in Vietnam

FOREWORD

In the summer of 1966, I was commanding a rifle company operating just south of the Demilitarized Zone (DMZ) separating North and South Vietnam. I received a radio transmission that a 'special person' was going to be joining my company for a week.

I was not given the name of the 'special person' due to security concerns. The very next day, a helicopter arrived at a landing zone carved out of the jungle and the 'special person' climbed off the chopper. I immediately recognized who this visitor was—particularly since he was wearing a distinctive eye patch over his left eye (from an injury Dayan received as a young soldier).

For the next week, General Moshe Dayan operated with my company. He went on patrols, sat in ambush sites, interviewed my Marines and served as a wonderful reference on questions covering war, fighting and geo-politics.

Every night we would eat together, and he would pull out a jar of Vaseline and rub some of it into his wounded eye to keep the scar tissue from tightening. He would do this without fanfare or embarrassment—just very matter of fact. By the end of his stay, we had become close—brothers-in-arms—based upon our mutual experiences under fire.

At the end of the week, my company was ordered to move to a location where it could be helicoptered back to a more secure location. We had to move some 10,000 meters through very rough terrain that included several fast-running rivers and streams.

As we crossed one of the streams, General Dayan lost his footing and slipped and fell into the rushing water. He went completely under but was able to regain his footing and stand up. As he arose from the river, water was pouring off him—he was a sight! Instead of being concerned,

General Dayan broke out laughing—at himself. At that moment, he won the hearts of all 150 Marines in my Company!

In 1998, as Commandant of the Marine Corps, I made my fourth visit to Israel. During that visit, I was given a book by the Israel Defense Forces—a book written by General Dayan. In that book was a description of some of his time with my company and pictures taken during the time he operated with us. I cherish two of the pictures. One is with the General and myself when he was coming off the helicopter on day 1 and another is a picture of the General crossing the stream just before he slipped!!

General Charles C. Krulak, USMC (ret.)
31st Commandant of the Marine Corps

PUBLISHER'S NOTE

This translation project has been many years in the making. Its purpose is to give English readers a unique work of personal observations of the American experience in the Vietnam War by one of the historic military leaders of the twentieth century. Awareness of the original Hebrew edition was made to the publisher by longtime military book dealer and collector George Arrington. His encouragement and friendship has been critical to the publication of the present edition.

Stone Tower Press gratefully acknowledges the assistance of Professor Al Bergstrom, Colonel, U.S. Army (ret.) of the U.S. Naval War College for his meticulous reading of the translation and helpful insights and corrections. As a combat veteran of the Vietnam War during the same era, his knowledge of the general and specific contexts of Dayan's observations is extraordinary.

Special thanks also to The Rev. Canon Samir J. Habiby, CDR, CHC, USN (ret.) for his assistance with gaining permission from General Dayan's wife, Mrs. Rachel Dayan, as well as his reminiscences of the Vietnam War based upon his two combat tours there. Naval War College Professor Emeritus Porter A. Halyburton, CDR, USN (ret.) and an American prisoner of war in Vietnam for almost 8 years (1965–1973), graciously answered questions about early US military aviation and countermeasure capabilities. Dr. Gina G. Palmer assisted greatly in the early stages of the book. Mr. Amnon Epstein was extremely helpful in direct communications with the late Mrs. Dayan in Israel.

TRANSLATOR'S PREFACE

Living in South Africa, the country of my birth. the Vietnam War was, for me, an abstract concept. Every couple of nights on the evening news, the numbers of casualties were regularly read out, but these seemed far away and of little direct interest. It was only when I immigrated to the United States in 1978, a few years after the end of that war, that I began to understand the seminal impact that it has had not only on the military, but also on the entire socio-political and economic system of the country. The effects of this war and its influence on military and political thinking are so vast that they can scarcely be exaggerated. They are still with us today.

By contrast, the State of Israel was a concrete reality for all Diaspora Jews, and we followed the developments there with keenest interest. Zionism was a key element of South African Jewish existence. I was too young to remember the War of Independence. but the Sinai Campaign, the Six-Day and Yom Kippur Wars and the attacks that occurred in between were key parts of our lives. Moshe Dayan was Commander of the Jerusalem Front during the 1948 Arab-Israel War, Chief of the General Staff of the Israel Defense Forces (1953–1958) during the 1956 Sinai Campaign, became the defense minister just before the Six-Day War, and personally oversaw the capture of East Jerusalem on 5–7 June, 1967, by which time he had become a figure of almost mythical proportions to Jews living outside Israel. Dayan shared the blame for Israel's lack of preparedness for the Yom Kippur War and resigned as Minister of Defense, but he played a significant role in negotiating a peace treaty between Egypt and Israel before dying in 1981.

It was therefore with the greatest surprise that I found out about Dayan's visit to Vietnam in 1966, just before the start of the Six-Day War. He was invited out as an observer by the Israeli newspaper *Ma'ariv* in 1966 and took the opportunity to expand it with visits to France, the United Kingdom, and the United States before, and Thailand and Burma after, his tour through the country. His contacts gave him access to the highest echelons in Europe and the United States. The book was published separately in 1977, after his autobiography *Avnei Derech* (Memoirs) a few years before his death.[1]

In France he meets with French Generals Lucien, Nicot, Koenig, and De Castries, and is wined and dined by various French, American, and Israeli notables. Despite their own failure at Diem Bien Phu, the French have a low opinion of American staying power and how they are conducting the war. Unless the Americans remain in Vietnam, they will not gain the support of civilians Despite De Gaulle's long-term vision, France is not strong enough to provide any significant assistance after their prior defeat. The only way they could have won the war was to slaughter hundreds of thousands of civilians and destroy their rice paddies—something that they were not prepared to do and would not have been supported by the French people.

Travelling on to London, Dayan meets with Field Marshal Montgomery in his country home. They take tea and Montgomery lectures him on various subjects. The American war in Vietnam is 'insane.' The United States must recognize China as a global power and allow the Vietnamese to decide their own future. He also opines on British mishandling of the Suez Crisis, and Eisenhower's decision not to enter Berlin in 1945.

Dayan continues to the United States, where he is granted a meeting with U Thant at the UN. U Thant opines that elections in Vietnam under the current system will be neither free nor fair. He has tried several times to contact Ho Chi Minh to bring the North Vietnamese to the bargaining table, but each time the Americans have declined to cooperate and continue their bombing campaign, and he now has no contact with Ho at all. U Thant has no problem with a unified Communist Vietnam. Dayan visits the Pentagon where he is briefed by two bragging, boastful colonels, who leave Dayan with many unanswered questions. He visits Walt Rostow in the State Department; it soon becomes apparent (even more so than before) that the United States situation in Vietnam is much more complicated than it seems, especially the issue of free and fair elections. A visit with the pro-war General Maxwell Taylor confirms this feeling. The issues of decreasing Viet Cong infiltration and convincing the North that they will pay dearly if the war is not stopped come up, also the use of smaller patrol and attack units. Dayan believes that bombing will not bring Ho Chi Minh to the bargaining table and wants to know why the Americans have not attacked the North Vietnamese during the monsoon season, when they could be lured out into the open and destroyed. Further discussions are held with other government officials as well as those in the private sector. Dayan spends an evening with Robert McNamara who promises to set up a meeting with General William Westmoreland when he reaches Vietnam.

He finally arrives in Saigon after a long journey with many stops. He is met by a public relations officer who gives him a one-sided, simplistic view of the war and he dines with the American press attaché. Dayan also dines with a high official from the South Vietnamese atomic energy board who tells him, in fear and the strictest confidence, that elections in South Vietnam will be neither free nor fair. He goes on a trip on a patrol boat, and it is immediately clear to him that no amount of inspection can uncover arms and ammunition concealed under piles of wood in civilian boats. He recommends that patrols be done by the South Vietnamese, with Americans acting as advisors. Visits to a US aircraft carrier and the USS *Constellation* follow. Dayan is greatly impressed at the size, scale, and smoothness of their operations but is still disturbed about the overall US war aim.

He visits a Marine unit in Da Nang, attends a briefing at headquarters, and goes on a platoon reconnaissance patrol. He wants to go out on another patrol with four to five men for 2 to 3 days in the field, but this is rejected. He visits a hospital ship and is impressed by its overabundance of supplies and personnel. He visits some villages: there are no men there (they are all fighting for the Viet Cong or the South Vietnamese army) and US soldiers help with everything. Dayan is very impressed by the US search and destroy missions. He visits the 1st Cavalry in Pleiku; they fly him to a fortified area where 700 mountain people (Montagnards) are working with the US in a special forces camp. Dayan is not impressed with this purely defensive posture. He finally gets his wish and goes on ambush patrol. He flies by helicopter to search for Viet Cong soldiers. They capture eight young men who have walked all the way from Hanoi to join the battle to drive the Americans from their land. They are simply dressed, with rubber sandals, minimal supplies and minimal weaponry. Dayan thinks: 'How can the Americans defeat such an enemy?'

Dayan visits a 'model farm' run by American volunteers. Although useful, philanthropy and assistance seem only useful for appearances sake. He visits a refugee and resettlement area and remains unimpressed. It isn't really resettlement, because work must be found outside. He visits a village on stilts defended with wooden stakes. The United States is trying its best but there is little fundamental change. In the meanwhile, he finds out that almost all the soldiers with whom he was supposed to go out on patrol have been killed. Dayan has many questions about how the 'search and destroy' missions are being conducted. Again, why not attack the Viet Cong during the monsoon season, let them concentrate in large units, and then destroy them? He visits the US base at Cam Ranh Bay and wonders who has given them permission to occupy the area in such large numbers and with so much equipment. Will they ever leave? At dinner he discusses the refugee problem. The Americans are missing a golden opportunity to resettle the refugees properly with expenditure of more money, and cannot find a way to defeat both Hanoi and the Viet Cong.

Dayan visits a Green Beret and Special Forces camp, with assistance given by Montagnards, mainly on the Cambodian border near the Ho Chi Minh Trail. The local military is unreliable. Dayan comments on the dangerous patrols under General Larsen. He also visits bases in coastal towns and another special forces camp. In the villages the Viet Cong take a great deal of rice from the people. They briefly occupy some special forces camps, take away a great deal of arms and ammunition, and then disappear. Roads are not safe for traffic (except for the area around Can Tho) other than convoys: the road network has died. Dayan makes a profound remark that 'The Americans are winning everything except the war.' He goes out on patrol again; the area is full of booby traps and mines, and one soldier is gruesomely killed when a sniper's bullet explodes a grenade in his ammunition belt. Dayan remarks that the higher-up members of the professional US Army 'celebrate this war because it allows them to practice their profession.' He visits South Vietnamese artillery units with American 'advisors,' and again doesn't understand why the US Army doesn't attack during the monsoon season.

He has a break in Saigon, meets with US Ambassador Henry Cabot Lodge, and a British press and military attaché. He then returns to the field and spends a significant amount of time with General William DePuy, whom he regards very highly as a commander and tactician. Troops continue to be killed by land mines, and no end of the war is in sight. Dayan dines with General Westmoreland but fails to detect a proper strategy to win the war in their discussions.

The last few days of his trip are taken up by visits to developmental and aid enterprises. He visits farms and inspects their irrigation systems and marketing strategies support by the American 'pacification' program. He regards inflation and the black market as the main problems. The US lacks a long-term plan. He visits Can Tho, a peaceful town with an intact road system. South Vietnamese control is stronger here than elsewhere. He visits the Mekong Delta region which is 50% under Viet Cong control. Markets are overflowing with all kinds of produce, but plowing is still by water buffalo and cows. Conditions in the refugee villages that he visits elsewhere are miserable—again, the Americans have no plan. Dayan travels via Long Xuyen to an area of Hoa Hao, a separatist Buddhist sect. Back in Saigon, General de Puy is depressed by a hard-fought battle with many American casualties. On hist last day in Vietnam, Dayan lunches with Chief of Staff and Minister of Defense General Nguyen Van Thieu, soon to become Prime Minister of South Vietnam. They both hold North Vietnamese General Giap in very high regard.

On the way home, Dayan stops off in Thailand and Burma. In Bangkok he consults with several high-level military officials. While they don't completely agree with American war policy in Vietnam, they are strongly anticommunist. He discusses with a local UN representative their plan for a TransAsian Highway and the opening of industry. Dayan meets with South Vietnamese General Duong Van Minh who is in exile in Bangkok. His opinion is that the Americans have taken too much of the fighting upon themselves. The Army of South Vietnam are trying to rule by force instead of with the people. Their leaders are corrupt and smuggling money out of the country prior to leaving. The war has broken the social structure of the country.

Dayan's description of his stay in Burma (Myanmar) is unique because at the time the country was almost impenetrable except for the press and diplomatic corps. Dayan arrives in Rangoon. There are fruit and vegetables in the markets and worshippers in the pagodas. But his hotel has 400 rooms and less than ten guests. He is successful only in meeting with the director of broadcasting, who is a major in the army. His efforts to contact the Department of External Affairs are fruitless. The stores are almost empty, but the people get free rice and vegetables. There is freedom of the press, but rule is by a government elite who stifle private enterprise.

He tries to buy a ticket to fly to Namsang, part of which is modelled on Israeli settlement methods with Israeli assistance. There is nobody on the scheduled flight, but he cannot buy a ticket without permission from the Department of External Affairs who will not speak to him. The Burmese way seems to be to do with as little foreign influence as possible. The people are adequately fed. Dayan meets with the North Vietnamese Ambassador with a view to obtaining a North Vietnamese visa and visiting Hanoi to judge the war from their vantage point. But a message arrives from the editors of *Ma'ariv* calling him home. He ignores invitations from

US Army Chief of Staff Johnson and Secretary of Defense McNamara, and flies straight home via Bangkok.

Dayan's views on the war in Vietnam come from observations, patrols, and long evenings of discussions, and his opinions are direct and to the point, both from a military and a civilian point of view. Even after all these years, they provide a unique point of view from someone with no direct military interest in the war. Israel had enough on its plate in 1966 without concerning itself with Vietnam. As befits a military man, Dayan's Hebrew is clear and crisp. He is fond of putting his thoughts down in logical, numbered and lettered lists. His acerbic, sometimes condescending style is well illustrated by references to 'californication,' the quality (or lack thereof) of British food, and some of the opinions of Field Marshal Montgomery and some American leaders. Many people from many countries are listed throughout the diary. Hebrew doesn't do well transliterated without vowels into names of people and places in French, Vietnamese, Thai, and Burmese, but I tried my best. Where I couldn't find reference to specific people cited, no footnote is given.

My grateful thanks to Hillel Halkin for assistance with translation problems and Tim Demy for unearthing this book and for suggesting that I translate it, also for his assistance during the manuscript and publication process, and also to General Charles Krulak (United States Marine Corp, retired) for his introductory comments, and all the people listed in the publisher's note for their valued assistance with the manuscript. The late Rachel Dayan owned the rights to the Hebrew edition and provided verbal permission for an English translation and publication, and my wife Addie bore with me throughout, as she always does. All errors are mine alone.

Peter C. Appelbaum
Land O'Lakes, FL
30 September, 2024

Endnotes

1 Moshe Dayan, *Avnei Derech* (Milestones). (Jerusalem: Idanim/D'vir Publishing Co., 1976); Idem., *Yoman Vietnam* (Tel Aviv: D'vir Publishing Co., 1977).

AUTHOR'S INTRODUCTION TO THE HEBREW EDITION

In 1966, I accepted the suggestion of the newspaper *Ma'ariv* to investigate the war in Vietnam. At the time, this war was, from both a military and political perspective, a central global issue. I was also of the opinion that there were lessons in it for us as well. During my time in Vietnam, I kept a diary consisting of direct impressions of my conversations with soldiers in the field, and what I saw and felt.

Because of its scope, I didn't think it right to include this diary in my book *Avnei Derech.*(*Memoirs*). But I was very happy to accede to D'vir's suggestion of publishing it as a separate book.

It is published here without changes, exactly as it was written.

M.D., January 1977

— 1 —

ON THE WAY

Paris, July 4, 1966

I am travelling at the request of *Ma'ariv* to report on the war. The first day doesn't go badly at all: I am staying at the Royal Monceau. Lunch at *La Duine*—as benefits princes and generals. I am invited for dinner by the editor of my book on the Sinai Campaign.

I have decided to travel via Paris, London and Washington, to hear political and military assessments and evaluations from French, British, and American points of view. Obviously, each country has its own, sometimes contradictory, opinions. I know of no previous war which has been accompanied at every step by so much internal and external criticism. This public war is just as powerful as the military one. At every step, the cemetery of military and public opinion accompanies the cemetery for those killed in battle.

The French foreign secretary cancels our meeting at the last minute because of 'unexpected workload.' I suspect that he took counsel with others—or perhaps with himself—and thought it safer not to meet with me.

Two other meetings take place as scheduled: a meeting with General Nicot, in his office,[1] followed by lunch with General Lucien.

These two generals have served in the air force in Indochina: Lucien, now retired, was an officer attached to operational headquarters, and afterwards served in political capacities, as French military attaché, and after that member of the French military delegation in Vietnam.

Despite my efforts to obtain his military assessments of why France lost the war in Vietnam, and of the current American effort, he concentrates on politics. He believes that the North is tired of war, and that I will be too late: By the time I arrive in Saigon, there would already be a cease-fire between the Viet Cong and the Americans. The American problem is to overcome negative world public opinion. (He believes that anyone who takes on the responsibility of war is forced to oppose public opinion). To overcome the North, it is necessary to destroy their rice paddies from the air. When I ask him if military victory is necessary for the prestige of America alone, or that of the international community, he replies unhesitatingly that it is of crucial international political importance. If the Americans lose the war and abandon Vietnam, the West will lose influence over the whole of Southern Asia. If free elections are allowed, he believes that South Vietnam, not the Viet Cong, would obtain a large majority of votes. If they are not preoccupied with the Viet Cong, the South will willingly accept the influence of the West—France, America. In summary, were it not for international and internal American public opinion—which interfere with proper prosecution of the war—everything could be brought to a peaceful conclusion (by military means, of course).

When the discussion (and the meal) ends, Lucien repeats his request that everything we spoke of should remain strictly confidential, otherwise he fears complications. I give him my promise, feeling that it wouldn't make a scrap of difference either way.

My meeting with Nicot is completely different. Even though he had spent 5 years in prison (due to participation in a failed putsch against de Gaulle), he retains his alertness and ability. At first, we exchange stories of the 'good old days' when France collaborated with Israel during the Suez-Sinai Campaign.[2] He speaks with bitterness about his incarceration and smilingly asks Yitzhak Israel (a member of the Jewish Observer Editorial Board who accompanies me) whether they have forgotten to change his newspaper address from prison to his new office. It soon becomes apparent that he is well-informed about Vietnam and follows what is happening there carefully. He is of the definite opinion that the Americans are not prosecuting the war properly. Their objective is wrong, and they are trying to achieve it by incorrect means. Their heavy bombings of the jungle are—and will remain—off-target because the Vietnamese units do not have sufficient information on enemy troop concentrations. He believes that the Americans should establish a network of small units of five to seven soldiers under American command. These small units should dog the heels of the Viet Cong: American bombers should only be called into action when significant troop concentrations are revealed. At the same time, army helicopters should be called in to surround and attack the enemy. By the time the Americans attack, they are usually 'running on empty.'

Nicot does not think that it is possible to prevent the North from infiltrating the South by bombardment, or any other means. They can bypass all obstructions, avoid any closures, simply by opening new roads, and melting into the jungles of the South. The only thing that can be achieved is prevention of concentrations of large units (brigades, battalions, divisions) with serious attack potential.

A country with such a military policy must be prepared for a prolonged war—many years of inimical 'co-existence,' not a radical action that ends the campaign 'once and for all.' He believes that the French did not lose their 1954 war because of lack of equipment or manpower, but because of incorrect military leadership. Dien Bien Phu was a tactical error, which occurs when a war is conducted with an unrealistic strategy that doesn't properly take facts on the ground into consideration.[3]

He has no clear answer to the wider political aspect of the war. On one hand, had the Americans evacuated Vietnam and allowed Ho Chi Minh to lead a united country—Vietnam would not have permitted Chinese authority, and would have kept its independence as a socialist (? Communist) regime.[4]

On the other hand, one cannot ignore its influence on neighboring countries (Thailand, Cambodia, Laos, etc.): Western influence would cease, and it would lose its grip on all these countries. France's past, and

America's present, failure would remove the West from South-East Asia. He (Nicot) does not believe in the stability of governments supported by America in Vietnam. This is the rule: 'East of Pakistan,' it is possible to govern with three things:

1. Money, to buy the leadership;
2. Intelligence, which France had but America does not have;
3. Military force, properly used.

However, the fundamental issue is that Asians do not want foreign domination, including local rulers imposed on them by foreigners. For this reason, their only concern is their own self-interest. If there would be free and fair elections in Vietnam now, they would choose the southern political arm of the Viet Cong—the National Liberation Front.

He also believes that the Vietnamese are not interested in modernization or American technology. They are interested in an improved standard of living, but with their own traditions. They have beautiful women and their own customs and are not interested in turning into 'californicators.'[5]

Nicot accompanies his words with maps and drawings of Vietnam, indicating transportation roads, rivers, location of units, helicopters, airplanes, infantry, etc. As agreed upon, when we part, I ask for his drawings. He signs them, writing his motto at the top of the page: 'The military problem in Vietnam is to capture the enemy.'

On our way out, he tells me that I am endangering my political standing by going to Vietnam: everyone will criticize and hurl garbage on me—Americans, Chinese, those who support the war and those who oppose it. From every side, they will constantly find fault with what I am doing. Apart from that, he believes that a visit there would be of no use at all, because I would not be able to observe the enemy, nor the war itself. I answer him that at least I will be able to see that I cannot see—even this would be useful.

I would have liked to suggest that Nicot visit Israel, but he has no spare money, and my standing is not such that I can invite him on Israel's behalf.

Evening (8.00 pm)

Dinner in the editor's home. Mocha Limon,[6] myself, General Koenig,[7] Massenet from Pompidou's office,[8] the editor, and his daughter.

Most of the table talk revolves around Israel, archaeology, de Gaulle, France's place amongst the world powers, etc. Naturally, Vietnam also comes up. According to my dinner companions, de Gaulle's attitude to Vietnam is based on two principles—Firstly, he is favor of national territorial integrity and opposed to fragmentation. Therefore, he is against division of Vietnam into North and South, etc.—Secondly, he is in favor of geographic continuity: Vietnam cannot be governed from California. Therefore, we must accept the fact that Asians must govern their own affairs.

I ask what the French government's opinion is: If the American war effort fails, will it lead to loss only of American prestige, or to collapse of the West in the area? I don't get a straight answer but conclude from what they say that de Gaulle knows the Americans won't leave Vietnam, so the question is only theoretical. De Gaulle's stance is central, and long-range.

A French delegation is currently in Hanoi, in an attempt by de Gaulle to successfully negotiate peace at a critical hour. He himself will soon be in Cambodia, and the head of the delegation—who was previously Minister of the Free French Forces—knows Ho Chi Minh personally, having parachuted in during the war to help the Vietnamese in their struggle against Japan.

After dinner (11.00 pm), we visit Shmuel Safir (Chichik), from the Israeli military delegation in Paris. I sit and talk until after midnight with him and his guest, Colonel Furcot (head of intelligence for the French Air Force).

The colonel gives the impression of being an intellectual. He shrewdly and clearly perceives the tangled problems surrounding the Vietnam issue. He is very skeptical of any solution, but at the same time does not pour scorn on a Western catastrophe—If the Americans fail, the main problem will remain public opinion. There was a time when such a rebellion would have been put down by killing of a couple of hundred thousand people. But today, Thank God, those bad old days are over. He does not believe that France was strong enough to succeed in Indochina. The problem was not lack of equipment or weak military leadership, but lack of a much bigger army which would have been necessary to overcome the North. Additionally, French public opinion would have been against large-scale conscription for service in Vietnam. The only way to force the issue, would have been for the South to invade and kill as many North Vietnamese as they could. After such large-scale slaughter, the remaining population would have been discouraged and beaten down. But the way things turned out the French were not able to defeat Vietnam.

The colonel and General Koenig emphasize that France today, a second-rate power, could not conduct such a war. De Gaulle is a great actor, whose rhetoric could make an impression seem like reality. But he doesn't have the necessary strength behind him to force issues. At the end of the day, the United States and Soviet Union will not permit a second-class power to start wars with international repercussions.

5 July 1966
11.00 am.

Meeting with the Minister of Fisheries. I suspect that our program for a joint tuna fishing enterprise in Djibouti will not become reality. The minister and his entourage appear surprised when I say that the costs of research—half a million dollars a year—would be divided equally between France and Israel. They probably thought that all they would have to supply would be experts.

My meeting with General Gallois is the most instructive thus far in military political theory.[9] We meet at his home for lunch together with Mocha Limon and Yosef Golan, the Israeli Military Attaché in Paris. Gallois monopolizes the conversation, so that I am hard put to squeeze a word or comment in. His doctrine is clear and ordered (at least to himself), and he has an answer to every question before it is asked. He has lived with, thought, read, argued, and weighed these issues for many years. His vanity is tickled that someone like me happens to visit him, and he 'practices' his theories on me.

He sees two basic errors in America's military policy in Vietnam:

1. Declared policy. America states that it is only in Vietnam to assist, advise and provide reinforcements. As soon as the campaign is over, they will leave. After all they have no interest in bases in Vietnam, territorial conquests, etc. General Gallois asserts that the population remains faithful to the Viet Cong. They, who remain after the Americans have left, will be the permanent rulers in Vietnam. The population understand that, straight after the Americans leave, the Viet Cong will settle accounts with all collaborators and will punish all who denied them aid.

   The guerilla fighters are either 'national liberators,' or 'criminal gangs,' depending on who governs the country permanently. The American stance, presenting themselves as foreigners and temporary guests, gives the Viet Cong legitimacy as 'national liberators:' this, at exactly the time when the population should be convinced to hand them over to the government as criminals. But they can only do this if they see America as their permanent ruler. I interrupt General Gallois, saying that he is oversimplifying the issue presenting the population's relationship with the Viet Cong as a reaction to the Americans' written declaration. I understand that the attitude of the population—if it is positive for the moment—to the Viet Cong fighters will change, one way or the other, according to McNamara's declaration.[10]

2. The second issue is that of American foreign policy. The United States has three possibilities:

    a. Isolationism, not to become involved in foreign issues;

    b. Reliance upon allies;

    c. A pervasive international presence. The United States is adopting the third approach, despite clearly having no imperialistic ambitions. If America wishes to rely on its allies (the correct policy, in General Gaullois's opinion), it should treat them with greater consideration and equality and encourage instead of opposing them in development of nuclear weapons.

According to America's present foreign policy, nuclear weapons are the exclusive property of two groups of nations. The first group is the United States and the Soviet Union; the second is Britain and France, both capable of defending themselves with nuclear weapons, if they do not endanger the United States and the Soviet Union. (China is near to belonging to the nuclear club).

America has decreed—deliberately or accidentally—that it should become an imperialist power and extend its influence wherever there is danger of the spread of communism.

General Gallois believes that America should state its goals openly and announce that it has no intention of leaving Vietnam. On the contrary, they wish to establish nuclear bases there: by remaining in the region, and coming to an agreement with Ho Chi Minh, who would become the head of state. American interests are global and security-based—they neither could, nor wish to, influence Vietnam's government. He believes all this to be achievable.

Gallois realizes that the United States could only proclaim such a policy if one of two conditions were met: that China would enter the war, or else that a new American President would be elected, with a new and different political platform.

Gallois's approach to the Vietnam issue is a part of his grasp of world military possibilities. I ask Mocha to find out whether Gallois's points of view have been published, because I have not made notes while he was speaking. As I remember it, the main points, etched in my memory, are as follows:

1. Nuclear weapons are a weapon of last resort in war. 'Marginal' wars must be dealt with by conventional weapons.
2. The United States and the Soviet Union will not allow other nations to start wars with international repercussions, so these smaller nations will be responsible for 'marginal' conflicts.
3. In the past, all methods of fighting, such as weapons and troop strength (divisions, force concentrations, airport availability, etc.), were vulnerable. War aims and expected military gains were, to a degree, safeguarded by distance. Today, the opposite applies—weapons such as submarine-based Polaris missiles, and land-based missiles that can be hidden and protected in hardened silos, are not vulnerable. However, civilian targets—population centers, entire countries, etc.—are vulnerable to destruction. This gives the possessor of such weapons enormous power: their dominance is unlimited, and they can easily be defended.
4. The make-up of military power has changed. Therefore,
    a. Population size has no more value. For example, Napoleon understood and made use of the fact that France alone could muster more divisions than all other European countries

combined. Today, a small number of soldiers suffices, relative to the power of its nuclear weapons;

b. A country's geographic (geopolitical) location is not permanent. Missiles can cross oceans, which previously protected far-off lands;

c. Raw material for steel manufacture was necessary in the past for weapons manufacture. Now, the quantity of steel is not the determining factor;

d. Instead of international bases and communications, space-based nuclear weapons are now possible, and space is open to everyone.

5. This development, created mainly by nuclear weapons experience of America and the Soviet Union, has removed the prior objective advantages of both nations' military installations. Theoretically, any nation can now increase its power, limited only by scientific development.

6. Like his predecessors, General Gallois does not believe that France—or any other nation—should overtake America or the Soviet Union militarily. No matter what happens, the financial capacity of the two great powers is too great, compared to other countries. General Gallois speaks with total contempt about the military and political leadership of the United States, and their military policies.

General Gallois continues his criticism of American tactical-military development, for example:

1. In Vietnam, in the beginning American airplane runways were too short. The Viet Cong attacked them with machine guns, and in response the Americans widened their defense perimeter. The

Viet Cong then started to use mortars, so the Americans widened their peripheries even more—and the net result is that they now finally have space for long runways, necessary from the beginning for all jet planes.

2. In NATO. America has created permanent operational centers with refueling facilities (he mentions a few). But their railway crossings, roads, communications, buildings, etc.—which serve as bases for troop concentrations—are ideal targets for an atomic strike. Therefore, as soon as a battle starts, they would be useless. NATO does not have comparable weapons for an attack on similar targets in the Soviet Union. The physical structure of NATO bases is therefore useless.

3. Provision logistics in Vietnam: A sailboat containing 3 tons of rice and ammunition is sufficient for 3 months for one company of Viet Cong. By contrast, a ship with ten times the tonnage of American supplies will be enough for only ¼ of the time. The logistical strength necessary for guerilla fighters is negligible.

At 4.30 pm I meet with General de Castries, the commander at Dien Bien Phu when it surrendered.[11] He tells me about the 'terrors of the jungle,' and the impossibility of an army fighting against the Viet Cong guerillas with conventional military means. He ridicules American speeches about industrialization of Vietnam, but also the suggestion that its population would all become Communists. Vietnam is an agricultural country; farmers who make their living from family-owned farms are not good material for rule by the Communist proletariat. At dinner (held by Mocha), I meet Bloch (from the Bloch Dassault family, manufacturers of French airplanes),[12] and the Minister of Free French Forces. I understand that France's official viewpoint is that the American conflict in Vietnam is a failure, and it would not be an international catastrophe if they lost the war and left the country.

6 July 1966

Meeting with Tran Van Huu,[13] prime minister of Vietnam from 1950 to 1952. His opinion is that American should stop the war and leave Vietnam. It can (or should) be replaced, as a first and formal step, by a multinational non-military force. Vietnam doesn't need foreign powers: the Vietnamese will rule themselves. They are not Communists, neither are they subject to China. Relations with France are still strong, and economic, cultural, industrial and trade ties can be strengthened and developed further—But only as sovereign and equal partners and not subordinate to France (this was France's error in the past). He does not believe that Vietnam will be conquered or imposed upon by China. This happened long ago, and the Vietnamese continuously rebelled against it.

Noon (1.15 pm). Lunch with Ambassador Walter Eytan:[14] I find a letter from him waiting for me when I arrive at my hotel. I suspect that he is complicating his life a little with the foreign ministry in Jerusalem, which doesn't want to legitimize my Vietnam trip.

Lunch at the ambassador's home (the ambassador, Yael Varad, David Scheinborn[15] and myself). It turns out that France maintains diplomatic relations only with North Vietnam. There is an 'economic delegation' from the North in Paris which fulfils the role of an embassy. Of course, there is a French representative in Hanoi. By contrast, there is no South Vietnamese representation in Paris, because France does not recognize the Saigon government, and there is no French representation in Saigon.

This fact, combined with France's recognition of Communist China, its most recent attitude to the Soviet Union, and France's recent exit from the military arm of NATO, are indicative of its political direction.

At 4.30 pm, I meet with General de Seguins Pazzis[16] (deputy commanding officer of the Paris *école supériore de la guerre*): a very intelligent man. His answers to three questions are as follows:

1. Will America's exit from Vietnam just be an expensive loss for them, or a catastrophe for the West in general?—It will not be

a catastrophe. He does not believe that the problem is competition between the influence of Western democracies and communism, but rather internal competition. Vietnam will not become a Communist country in the classic sense of the word. However, there is no way to prevent influence from China and other Asian countries. Spheres of influence must not be seen as ideological, but rather as geographic: nations, cultures, etc.

2. Are the Americans conducting the campaign correctly from a technical-military point of view?—The problem is not military: it is that the Americans are regarded as foreigners, and the Viet Cong as patriots. American popularity is lower now than it was 2 years ago.
3. Could France have won the war in the 1950s? (The general was in Vietnam for 5 years)—Yes, not by piling up more arms and ammunition, but by a policy of liberality and equality with the Vietnamese. General Leclercq[17] was of the latter opinion, but the French civilian government initially refused, and afterwards it was too late.

Dinner with Koenig: he Mocha, myself, and others. Wide-ranging discussion. Regarding Vietnam:

1. Around 1954, Koenig was offered command of the French campaign in Vietnam. He set out five conditions—four political, the fifth (really the first) military. Two French divisions would be necessary to win the war: total French entry into the war, with equal participation of soldiers from other countries under French rule. The government refused.
2. In all my conversation with Koenig, he repeatedly emphasizes the fact that de Gaulle creates the impression (for himself and others) of the greatness of France. However, this is not based on fact

because France today is only a medium-sized power. Koenig sees great danger in the illusion of power. According to him, the same false impression of France's military power was created between the wars—When the real test of strength came, the legend shattered into a thousand pieces, and the people lost faith. De Gaulle is a great actor in the fullest sense of the word, who gives the impression that France is great and powerful; however, this is incorrect, and the country fails when tested. Koenig sees no way in which France can regain her former greatness.

After dinner, we join a gathering with the feuilletonist Art Buchwald.[18] 'The entire world' (mostly Jews) are represented at the party: Raul Levy, Kaplan (Vietnam Press), D. Scheinborn, Elie Rothschild and Irwin Shaw.[19]

Summary of my Paris Impressions

1. Regarding the French failure in Indochina:

   a. The root of the problem was not military, but socio-political. The French made the mistake of not treating the Vietnamese as equals, as a real sovereign nation. As a result, the population saw the French as foreign invaders, and the only way the Vietnamese could have been conquered would have been with great brutality, and impossible and unimaginable mass slaughter. An example of inequality: both the Vietnamese and the French were graduates of the same French school system. However, even if the best student in the class was Vietnamese, the French student received double the salary, and a much better government job.

   b. The French assumption that establishment of order in Vietnam could be based upon fundamental differences between Arabs from North Africa—or Africans living in French colonies—and Asians (Cambodia, Laos and Vietnam). Asians who grew up with French culture would be prepared to cooperate closely with France (based upon sovereignty and equality). *Unlike* Arabs, they have no nationalists who don't want foreigners in their country. According to the French, French schools are still full to the brim of Vietnamese students. Asians are not xenophobes and racists, at least not against the French. (They are of the opinion that there is no possibility of direct, unobstructed, uncomplicated contact between the Vietnamese and the Americans, who are newcomers with a foreign culture. The French have been in Vietnam for 90 years—three generations!).

c. As to France's military failure, first France did not make the necessary effort. They didn't call up the reserves, or men liable for compulsory military training. Only the regular army fought in Indochina (as was the case in Algeria), and this did not suffice. This led to the fixed impression—amongst both the fighting French and the Vietnamese rebels—that France would not exsanguinate herself in this cause. This weakened the soldiers and encouraged the rebels. Therefore, General Koenig made his acceptance of command conditional on the calling up of two divisions of *petits français* (young men born in France).

d. Lack of all kinds of equipment, general and specific. Emphasis on incorrect tactical and strategic leadership. The previous French commander presented Dien Bien Phu to De Castries as a type of 'mountain cavalry stronghold.' He explained that, from this base, the French would be able to control the entire area. In fact, it was a trap: The Viet Cong positioned themselves on the surrounding mountains, well dug in and camouflaged.[20]

De Castries maintains that the task given to the soldiers in Dien Bien Phu should have been given to the air force, because the task was unsuitable for infantry.

e. Incorrect assumptions of what could be accomplished militarily. It was impossible to prevent infiltration, and aerial bombardment could never destroy the Vietnamese guerillas. The typical Vietnamese (now Viet Cong) guerilla's equipment consisted of a bandolier with a bag of rice, a rifle or machine gun for short-range hand-to-hand combat, and a ditch or tunnel, in which he could quickly hide, safe from aerial bombardments. Therefore: it was a mistake to try and clean out the country of Vietnamese fighters. The French Army should have concentrated on essential areas, from the time when it took over administration from the United Nations.

2. As far as the Americans are concerned, comparable points can be made:

   a. The French do not believe in an American victory; but they will not be beaten either (from a military point of view). They do not think that Ho Chi Minh will be forced to the bargaining table with the Americans, or that they will be able to convince the people to support them against the North.

   b. Most people do not foresee any catastrophic consequences if America loses the war or leaves Vietnam. After a period, a government will be formed in Vietnam which is not Communist in the classic sense of the word, and not subject to China. The government will be independent, capable of developing friendly economic, cultural and other ties. General Gallois disagrees: he sees the world as divided between the United States and the Soviet Union, the only two global powers.

c. Although they recognize the power of the American military, they don't believe that the bombings will be effective or fulfil their aims. Most explode harmlessly in the jungle, without harming the Viet Cong.

d. The French are particularly contemptuous of 'aid to South Vietnam against the Northern invaders.' The great majority of Vietnamese people want peaceful lives and cessation of hostilities (which have already continued for 25 years). They have no faith in the current government or in any other government supported by America, or one that comes to power by military means.

e. The French believe that the American mind-set of technological modernization, and culture in general, are foreign, and cannot win the hearts and minds of the Vietnamese people.

In summary, what would the French advise the Americans to do now? I would say simply: stop the war, and concentrate in several autonomous bases, to:

1. Protect global American security interests (nuclear bases), not create the impression of military failure (not expected from the American Army).
2. Allow the Vietnamese to form a government of their own choice (obviously—North Vietnamese) and keep order with help from the United Nations.
3. Renew Vietnam's relations with the West (the French suggest the aid of a multinational force, in other words with French cooperation) based on equality and sovereignty, and the fact that Vietnam does not want to be subject to China.

London: Conversation with Montgomery[21] 10 July 1966

At 12.30 pm, we travel to the 'Old Mill,' Montgomery's country home. Ambassador Aharon Remez[22] joins me for lunch, which is a third of the price of one in London, and seven times tastier. We arrive at Montgomery's home at 3.15 pm. He comes out to meet us, shakes hands, and, without preliminary pleasantries, asks what is happening in the world. I say that I have come to ask him exactly that. He is prepared for the question and naturally responds: the world is going mad—referring to Americans fighting in Vietnam.

When we finish the discussion 2 h[ours] later, he says to me: 'come back to visit when you return from Vietnam and tell me what you think: tell the Americans they are insane.'

My book sticks out from others in his entrance bookshelf. I'm not sure whether this was because of its blue color, or for my own benefit.

He has certainly gotten older but looks much better than when we last met in the House of Lords. This time, he rejoices and despairs, speaks and laughs, incessantly; his thoughts and descriptions are amazingly clear, and he remembers dates, societies, even Hebrew place names. He remembers his visit to Petra, and after that to Mount Pisgah (Nebo), from where Moses our Teacher looked out on the Land of Israel but could not enter it. He pronounces the word *Pisgah* in Hebrew.

At 4.30 pm—in the best British tradition—the gong sounds. We go down from Montgomery's study on the first floor to the dining room on the ground floor—to take tea. Walls in both rooms are covered with photographs of Montgomery in uniform at different time periods, and other leaders and commanders, such as Tito,[23] Churchill, etc. Also, landscapes painted by Churchill himself.

Vietnam is the main topic of conversation. Montgomery has a clear opinion on the issue itself, and how to solve it. In essence—He believes we must recognize that China will be the dominant Asian power (he emphasizes this). In 20–25 years' time, China will have a population of a billion, and also nuclear weapons. (For some reason, this was the only time that nuclear weapons were mentioned during our conversation. Neither of us touched on the question and influence of nuclear weapons on the other issues). Therefore, China's hegemony in this region must be recognized. The West (in other words, the 'white man') must organize its bases, including in Australia and India. Montgomery believes it would be possible to establish special ties with India, and that India would not enter China's sphere of influence. From a military point of view, he says, with outward humor but serious meaning: 'There are laws in war that must be adhered to. Firstly—don't even think of Moscow: it has no reason, and no chance of success. Secondly—don't make war on China using infantry: it also has no chance.'

China's influence in the region must be accepted even if it is a Communist one. What is wrong with that? He believes that this type of regime suits Asia, where ½ of the world's population currently live. But it isn't suitable for the West. The Americans err in their wish to confer their 'way of life' to others.

The borderline that differentiates the Asian-Communist-Chinese world from the West, is the Indian Ocean coast. Taken that way, the Middle East belongs to the West. (I don't think that he used the words 'white man' as a sign of superiority, just as an accepted figure of speech). The Chinese are not a nation of seafarers, and the West must strengthen its hold on the oceans.—'3/5 of the world is ocean'—and be present with suitable strength on its borders.

Russia can and should be brought closer to the West—that is, from a military point of view, the United States, England and France. Russia belongs with the West. On one issue he refuses to budge: the division of Germany. Russian concerns and suspicions are deep and well-founded, and they will never, under any circumstances, allow Germany to become a nuclear power.

Before anything else is considered, America must define their strategic policy: they are currently not doing this. He had spoken with Nixon:

'I asked him: what kind of agreement do you wish to reach? After he had orated non-stop for 20 minutes, I had to ask him the same question again, because he didn't answer me the first time. They do not have an answer.' The first question to be answered, is the one he posed: Is it agreed upon, that the Chinese will have Asian hegemony for the foreseeable future?

All other steps should be organized on this first basic assumption. Asia must not be vacated in haste. That would be a grave political and economic error. It must be done gradually and methodically. All tactical issues must be in line with the strategic objective. What the Americans are doing today (he doesn't mean one specific step, such as the bombings), is insane. They are fighting without logic. It may be assumed that Vietnam wants to unify and become Communist, as is the case with Korea[24] and, at the end of the day, Cambodia and Laos (not in haste, and not immediately). In any case, anyone with another suggestion must come up with proper strategic objectives, according to which day-to-day events should be directed.

In his opinion, what the Americans are doing is neither good nor logical. They are conducting the war by military means only, without taking political issues into consideration.

The same thing happened during the Second World War.—He suggested to Eisenhower that the Allies advance to Berlin because this was the main political objective. (Wars are fought for political objectives). However, Eisenhower's approach was purely military progress—after military success, other issues could be considered.[25]

The problem today is leadership, in which Western leaders will believe and have faith. A real leader must be gifted with long-term vision. Churchill had it during the war, and de Gaulle has it today: These two men also have 'second sight,' and intuition. The late John F. Kennedy probably had it as well. But men like Johnson, Wilson[26] and other experts, work from day to day, with no strategic vision or objective.

Montgomery directs the discussion, which touches on many issues. He sometimes questions, but mainly—lectures. He is currently writing a book on the history of war—from man's earliest beginnings to the present day (about 400 pages).[27] Two young men are working with him (they join us for tea), doing the necessary literary research in the British Museum, 'where long-haired young men sit—that is not for me.' He recognizes that a significant part of his words are based on opinions of military-strategic problems in his coming book. He is now working on global issues, to arrive at a viewpoint encompassing the whole world as one unit, and history as a series of failures and successes, which can be analyzed and explained. This encourages him in a similar approach to concrete questions.

There is no doubt that his own life experiences—which are now in their final phase—have influenced his point of view. He repeats—frankly, not pretentiously—'This is what I have learned during 50 years as a soldier.'

There are two main lessons, which he repeats and emphasizes repeatedly: First, war aims must be fixed with clarity and accuracy: exactly what do we, ultimately, wish to achieve?'During the Suez Crisis (1956), I asked Eden[28] what he wanted to achieve. Eden replied that he wanted to throw Nasser[29] out. When I asked him what happened after that, he had no answer. That is no good: throwing Nasser out was not a strategic objective, only an intermediate aim.'

Second, conclusions must be drawn and aims achieved without doubt and hesitation. 'If we wanted to throw Nasser out, we would have had to occupy Cairo. During the Mandate, we fought against you, because you ruled the country illegally. I said to Bevin, who was then secretary of state for foreign affairs: "One of two things must happen: either we crush the Jewish rebellion, or we leave Palestine." It is impossible to fight with one hand tied behind the other. Bevin[30] replied that it was impossible to wage total war against the *Yishuv*.[31] Therefore—I said to him—we must leave Palestine. And we left. That is the way—either win, or withdraw to another place, where can make a stand. We must draw proper conclusions—and either do what must be done or leave.'

About the Egyptian Army:'I know them since El Alamein: I called them "10 minute warriors." After 10 minutes, their fighting spirit disappears.'

Wingate: 'He was unbalanced. When I was army commander in Palestine, he acted as though he had assumed the role. Field Marshal Slim couldn't stand him. The only decent thing Wingate did in his life was to die in an airplane crash.'[32]

'De Gaulle—is the greatest living leader today; the second is Churchill. He has vision and almost supernatural "second sight." He knows what he wants—and he wants the right things. He fulfils his objectives and, if someone disagrees with him, he tells him to go to hell.'

The Suez Campaign: I said to Eden: Keightley is the wrong choice for commander. He is a good man—but in war one only needs bad men, who can achieve victory.'[33] When Monty speaks of our current lack of world leadership, he says that England had lost a million men in the war. If they were living today, they would be in their fifties, and many would probably have been leaders. I suggest that maybe his two young assistants would become leaders. They both smile but Montgomery stares at them and says to their faces: 'These? No, not them.'

He asks how old I am, I say 51, He says that I was still young. 'When you were born, I had already killed Germans: obviously, not enough. But on the other hand. I did not attend the Nuremberg trials. Statesmen, not generals, should be put on trial. The task of generals is to win wars; they should not be put on trial and hanged if they lose.'

About himself—'I have been a soldier for 50 years. I rose through the ranks—from platoon to army commander—All my time was spent in the field. I know war.'

The words of a commander must be few, and concentrated on principal, not secondary, issues. But some details—which are the primary domain and responsibility of the commander—are essential for a successful campaign.

'When the war ended, I was without a home, because German bombs had destroyed it: not even a teaspoon remained. So, I bought this house (an old flour mill, which had been destroyed) with the little money I had. No one lived there, so I rebuilt it. Before, I had money but no house—now, I have a house but no money.'

When we speak about the Suez Campaign, I see that he has expert and detailed knowledge and has read the necessary books on the subject. When I mention something and want to go into it in detail, he interrupts me—'I know, that is in your book.' He remembers the exact date on which the Security Council agreed to a cease-fire (July 11,1956). He remarks that the British commander on the ground should have acted like Nelson by disregarding orders[34] and completing the campaign (by conquering the Canal Zone).

Throughout our visit and discussions, Montgomery is very cordial, smiles, and is full of opinions. When tea ends, he takes us for a tour of his garden and describes in minutest detail the names and seasons of all the flowers.

He speaks highly of Ben Gurion[35] and requests that he should come and visit, if Ben Gurion would be in London. Aharon Remez has invited him to Israel, but he answers that he was very busy writing, with little time for travel.

He relates that, while he was commander in Palestine the telephone rang, and someone said to his adjutant: 'It's the Irgun[36] speaking: we have a bomb for Montgomery.' The adjutant asked if he was joking, and the person answered angrily: 'No joke: There is another bomb—for you as well.'

11 July 1966

Lunch with Aharon and Lord Walston,[37] undersecretary of state for Asian and African affairs He was recently in Vietnam and is the only impressive individual whom I have met here. I want him to clarify England's position on Vietnam, and how they think this problem should be solved. We have a bad meal in an expensive restaurant, exclusively for members of parliament and their guests. Lord Walston makes a good impression—intelligent but not too serious or intense. Of course, it is impossible to gain a proper impression by a fleeting encounter.

In contrast to the official view, he doesn't see anything wrong in Americans bombing North Vietnam's fuel facilities near Hanoi. On the contrary, Americans must demonstrate strength and resolution. As for the overall solution—he believes that both the Americans and the North should strive to calm the situation down and diminish military activity. This would allow the Americans to leave Vietnam without paying too high a price. The world must make peace with a unified, Communist Vietnam (Viet Cong) without presence of American forces. This could be achieved, if it becomes clear to the North Vietnamese that they cannot overcome the Americans in open battle, and therefore act as if the Americans 'achieve' their goal—allowing them to leave the country. Then the Viet Cong would essentially become the government, if such a step is deemed necessary by both sides. The Viet Cong could torpedo this development if they continued their attacks, preventing the Americans from leaving without expenditure of blood and treasure; their withdrawal would then appear as if they were running away. The Americans could also torpedo this plan, by continuing their own attacks, thereby forcing the North to continue military activities, as a reaction to American strikes.

All this, of course, is on condition that both sides wish to end the war. However, there are no signs that the Americans genuinely want to leave Vietnam, or that the North are prepared to calm the situation down.

If the Americans leave Vietnam—he believes that they should remain in the area—in Thailand, Cambodia, Laos, etc.—and not leave East Asia completely.

This depends, quite obviously, on American withdrawal with or without obvious loss of blood and treasure. He believes this to be the position of Her Majesty's Government. Lord Walston says that, if accepted by the Americans, it could become England's official stance; however, without American acceptance, England cannot officially endorse it.

Tea at 4.30 pm with Zvika Zafriri (from our British defense delegation)[38] in the home of Julian Amery, conservative member of parliament.[39] A giant library in an impressive house.

Amery repeatedly expresses his low opinion of the United States. However, on this issue (unlike previous occasions), he accepts America's position. (He had visited Vietnam in 1962 and met with Ngo Dinh Diem[40] when he was prime minister of South Vietnam). He is of the opinion that America must continue until it has won the war—if necessary, even take Hanoi. The Chinese have serious problems feeding their huge population and are faced with three solutions: 1) conquer Siberia; 2) conquer Southeast Asia; 3) reach an agreement with the West, which will lead to economic assistance.

The Russians will prevent the first possibility, and the Americans the second, so they will be forced to turn to the third. He believes that the forces of South Vietnam, with American help, can conquer Hanoi. However, the Americans are faced with both external and internal problems; the latest bombings are directed more against Senator Fulbright[41] and Robert Kennedy than against the Viet Cong. The Americans will intensify the war, to present Americans who oppose it as unpatriotic. I told him that I do not agree with his theory. It is Ho Chi Minh who is waging this war, not China and, even if he wins, China will not acquire additional living space. I am also not convinced, from what I have heard until now, that South Vietnamese forces can conquer the North.

Shimon Peres[42] is much more impressed with Amery's opinions (which have been well thought out and researched) than I am. Amery says that he is travelling to Portugal for 3 months to write his memoirs, like all elder statesmen. (Who knows whether he will be able to return to political life after having lost the last parliamentary election)?

I tell Amery that I would like to meet with Lord Head,[43] the last High Commissioner of Malaya. (I had met Head in Nigeria when during its independence ceremonies: he was the British High Commissioner at the time).

Amery reaches Head by telephone in his home outside of London: it turns out that we will not be able to meet. I speak to him briefly over the telephone: he refers us to his speech in the House of Lords, during the debate on the Far East last May. I ask Zvika to obtain the minutes and send them to Israel. We agree to meet, if I return home the same way from Vietnam to London.

12 July 1966

I meet Gentrali from Rome in the hotel lounge, on his way to Moscow. I know that he has travelled to Vietnam: he agrees with those who don't like what is going on. Italian public opinion is against the Americans. In his opinion, time is working in favor of the Chinese. It is in their interests that the war continues; it also allows them to drive a wedge between Russia and the United States. If the Russians help the Viet Cong, they will become entangled with the Americans; if not, they will have problems with the Communist world, something they cannot allow.

At the United Nations
13 July 1966, New York

Meeting with U. Thant, UN Secretary General at 4.30 pm in his office.[44] I ask that Israel's UN representative, Michael Comay,[45] arrange the meeting: he accompanies me. The meeting lasts about 30 minutes, and only ends when U Thant's secretary enters, telling him that others (ambassadors from Cyprus and Turkey) are waiting to see him. The mere fact that U Thant agreed to meet me is a special tribute: he has been absent from New York for a long time and only returned last night. He has every right to say that he was too busy to meet me. He asks that the meeting not be publicized. Comay agrees with his secretary that, if the newspapers would enquire, they would say it was merely a courtesy call because we had met previously (U Thant had accompanied U Nu[46] on his visit to Israel).

Obviously, Vietnam is the subject of the discussion. Apart from a few questions which I pose, the conversation is a monologue by U Thant: appraisal and clarification of the situation, almost a kind of personal and political outpouring out the heart.

I ask him two questions: Firstly, the UN's role in this dispute; secondly, the method—unconnected with the UN—he believes should be adopted to end it.

U Thant does not believe that the UN should get involved in any part of Vietnam at the moment. Not only would it be a waste of time, but it might complicate other UN activities.

His main concern is the question of elections: he has requested UN supervision of the upcoming September elections in Vietnam.

UN supervision must not be fictitious—a kind of international approval, without being able to promise that the elections are free and fair.

Firstly, UN staff can obviously not control the elections, which will need many supervisors: about 2,000 ballot boxes will require c. 4,000 supervisors. Secondly—and this is of primary importance—the promise of free and fair elections does not begin at the ballot box. Each citizen must be promised the right to vote without coercion, and each person running for office must be able to present himself properly.

U Thant does not believe that such elections would be held in Vietnam. He reminds us of South Vietnamese bulletins published in anticipation of the elections: votes of anyone who supports the Communists, or anyone who asserts peace, neutrality, and unity, would be invalid. There is therefore no doubt that elections would not be free, and there would be no justification for UN approval by sending so many inspectors. U Thant speaks scathingly about General Ky:[47] he is certainly not the right man to be head of state. He mentions a certain general who is currently in exile in Bangkok and cannot return home because General Ky has made it impossible.

As to the second question, he believes that:

1. The bombings must stop (mutual agreement to a real cease-fire would be impossible and impractical, with so many small units scattered in the jungle).
2. All warring parties should be summoned to the bargaining table—the Viet Cong, Americans and South Vietnamese.

He doubts whether any of this would be possible at the present time: China's influence over Ho Chi Minh has increased during the past year, and China refuses any contact with the United States. The Viet Cong do not believe that the Americans intend to leave Vietnam in the foreseeable future: on the contrary, they are currently investing a hundred million dollars, just on one airfield.

He believes that the position is deteriorating:

1. Chinese influence and aid to the North with arms and industry is increasing.
2. There are about 20,000 Chinese advisors in the North: any harm to them during the bombings would hasten direct Chinese military involvement.
3. Other Communist countries, such as the Soviet Union, Poland and Romania have announced their agreement to send volunteers to help the Viet Cong.

U Thant does not hide his bitterness and scathing criticism of the Americans. The gist of his words is simple: they simply don't want peace. Some American generals (he does not name names) incite Johnson to attack (and bomb) China's nuclear facilities. He tells me that in September 1964 (about 2 years ago), Ho Chi Minh wanted to negotiate with the Americans, to try and end the war. U Thant had acted as mediator and informed the Americans that Ho Chi Minh was ready to send an authorized representative to negotiate. The Americans declined: first they delayed their response for 3 months (until after the American elections), then they said no. Negotiations would have been interpreted in South Vietnam as if America was betraying them, and 'selling them out.' Ho Chi Minh's agreement at that time was not connected under any circumstances with his unwillingness to become enslaved to Chinese influence. A year later, in May 1965, U Thant again acted as mediator. Ho Chi

Minh (U Thant knows him personally) was again prepared to negotiate, but this time with the condition, that the Americans would stop bombing the North. (Not temporarily, but permanently). The Americans refused. In August 1965 U Thant tried yet again to contact Ho Chi Minh, but he refused. From May 1965, there has therefore been no contact between U Thant and the Viet Cong.

In summary: U Thant sees that the Vietnam issue is becoming ever more serious; reasons he gives are:

1. Increased Chinese influence over the Viet Cong. The Chinese are more inflexible than the North, and not prepared to compromise with anything based upon Americans remaining in Vietnam.
2. Other Communist countries might send assistance to Vietnam in the form of men and weapons.
3. The government in Saigon is illegitimate.
4. Elections held there are not genuine.
5. The United States does not wish to reach a peace based on compromise.
6. The ultimate purpose of the United States is to remain in Vietnam.

Comay says that U Thant told him that if Burma would have requested American aid at that same time, it would have become a second Vietnam—a battlefield between the Americans and the Chinese.

When our discussion ends, I tell U Thant that I would like to meet with Ne Win.[48] He agrees with the idea: I ask him if he could relay his support to the Burmese ambassador in Washington: he says he would. I will try to obtain a Burmese visa in Washington.

Comay tells me that the United States only refers the Vietnam question to the various institutions of the UN when convenient, as a form of publicity. But in reality, America conducts its own foreign policy regardless of the UN, in important disputes such as Cuba, Vietnam etc. They turn to the UN with matters that don't particularly interest them, like Yemen, Pakistan, etc. Comay also explains U Thant's feelings. If I would have to classify our discussion it would be as the 'moment of truth.'

U Thant's opinion of the steps that must be taken now are as follows:

1. The Americans must diminish their military activities: this will, in his opinion, be met by a comparable reaction from the Viet Cong.
2. Change the government of South Vietnam.
3. Free and fair elections (in fact, although he didn't spell it out, this means a Communist government).
4. The Americans must depart, and in the meanwhile, show signs of doing so, and not vice versa, by building permanent bases.

Washington, DC
14 July 1966

I arrive in the afternoon and am met at the airport by the deputy Israeli military attaché (Major General Yosef Geva[49]), Lieutenant Colonel Amir (whom I don't think I have met previously).

First meeting at the Pentagon 1.00–4.00 pm.

My escort—Colonel Hayes—accompanies me to a briefing by two army spokesmen—Colonels Seidel and Miller: they explain to me the differences in command between the American and Viet Cong Armies, their relative strengths, and the structure and activity necessary to safeguard the safety of civilians. After their introduction, there is a question-and-answer session. The Viet Cong have a total of about 250,000 soldiers in the entire country—including civilians who assist them only in their own sectors, and political commissars. The South Vietnamese

Army has about 700,000 soldiers and the Americans 300,000, in other words a ratio of one to four in favor of the Americans. If we subtract the burdensome American administration and add their availability of air power (helicopters), the ratio becomes one to three. These numbers do not of course include Viet Cong forces from the North down to the seventeenth parallel.

The senior colonel of the two, Seidel, opens by apologizing, admitting that he is confused—How can a run-of-the-mill colonel in the American Army brief 'The acclaimed General Dayan?' However, it immediately becomes clear that he knows the answers to all questions, is even prepared to introduce new ones, if that is my wish ...

It is obvious that a map briefing in Washington doesn't clarify much, nor yield any new information. In effect, all that I hear today is what I have heard and read already. Answers to my questions aren't convincing either. One of the issues that requires explanation is the question of Viet Cong attacks during the winter (monsoon) season, which have thwarted General Westmoreland's[50] forces previously. I ask my hosts why the Viet Cong were not allowed to organize, concentrate, and attack—and then dealt with. The answer is that Westmoreland didn't believe he could take the risk of not being able, at the last minute. to overcome the Viet Cong Obviously I am not qualified to judge, but with a ratio of 1 to 4, and the North lacking (in the south) armor, artillery, and especially an air force, against the Americans who have all these at their disposal, it is difficult for me to understand why the Americans did not want to take the decisive step. They had a rare opportunity to deal the North a very heavy blow but were prevented from doing so because the North's forces were divided. Nevertheless, I still feel that this point requires clarification. Perhaps the North did not plan an attack? Perhaps they were convinced that they would not be able to overcome the Americans by a concentrated attack with large forces? Perhaps Ho Chi Minh—not the Americans—cancelled the plan of a large attack during the monsoon season?

While leaving the Pentagon, we pass their bookstore. Colonel Hayes—my escort—shows me that my book is on sale. I want to ask the clerk if many people are buying it but suspect not—and do not do so.

The military situation, victories, great technical power of the Americans, etc. are presented to me by the two colonels in a most unpleasant way: bragging, boastful, without intelligence. I do not oppose self-confidence, but the impression here is the personal self-satisfaction of a powerful man showing his wares off to guests.

I feel this way mainly when two issues are discussed—Firstly, when I ask them whether there had been any developments or changes in tactical approach during their 2–3 years of operations in Vietnam. In other words, did they not feel that certain tactical methods were unsuitable for that type of warfare, and should be modified or changed?

Their response is absolutely negative. Just the opposite: they feel that results were much better than expected. I want to react by saying: 'The American Army never makes a mistake!' But I remember that I am their guest.

The second issue is—maintaining the trust of the civilian population. The two men—rightly—lay great emphasis on this point. The goal is to gain the confidence of the Vietnamese population—scattered in small villages and isolated homes throughout the country—in the face of terror, 'tax collection' (rice, money, forced labor) and conscription by the Viet Cong.

My host Colonel Hayes explains that about half the population in the South (not in the field) have full confidence in their government. A quarter support the Viet Cong, with a quarter in-between. During the past year, the number of people living in areas under American control was about 10% less than this year. Their program and slogan-of-the-day was: 'full security during 1966!'

However, if one scratches the surface a little, removing formulas and statistics, it becomes obvious that the picture is far from simple or clear. Even in areas under American control, it is difficult to completely protect civilians against acts of terror. Even if a specific area has been thoroughly searched and cleaned out, and all enemies or enemy suspects imprisoned or killed, there is still no guarantee that, after the army departs, new Viet Cong forces will not penetrate and take control again.

I do not leave my hosts with a feeling of satisfaction. I realize that momentary impressions cannot properly measure a situation. This is especially the case when discussion only lasts for a short time and depends on personal style and facial expressions of two army officers. Truly, I do not come to any firm conclusions and only mention my reaction as I felt it at the time (quite obviously completely subjective). Perhaps this feeling is based on my past: in all past situations, even in the presence of victory, I have never really been convinced that it wasn't possible to do better. My military actions themselves have never been accompanied by 'joy in creation,' but mostly by just the opposite—a feeling of heaviness.

15 July 1966

I am accompanied throughout the journey by the 'Jewish problem.' First in Israel—opposition by the government, department of foreign affairs, etc. Then in France, England, and here as well. The embassies have received instructions not to cooperate with me and only help me if I ask as an ordinary citizen, for personal reasons. What has really happened is of course something different. The personal relations between myself and the ambassadors—Aharon Remez, even Walter Eitan—transcend official concerns, and it isn't possible for us to avoid meeting, talking, enjoying meals together, etc. Each country is convinced that my official address is the Israeli Ambassador, to whom they sent all requests for meetings, etc. For lack of an alternative, the embassy serves as a kind of communications office with me (mostly with the military legation, not the ambassador). The truth is that, when I need help (and this does happen), I turn to my friends, most of them in military legations and department of defense delegations. At the end of the day, the ambassador in France and especially here in the United States, guides me by whispering delicate issues about the Vietnam problem—as is seen in each country—in my ear. The main condition is that I do not say or write things that might harm Israel.

The climax of this business comes here. I know about these instructions to 'avoid' me unless I make a specific request—so, when I arrive, I telephone a member of our delegation, Afi Evron, and, after the usual pleasantries, tell him that 1) I don't need anything; 2) If the ambassador himself is interested in providing guidance, I would obviously be pleased to meet him.

Afi promises me an answer, but I don't hear anything from him. The only one who does appear is Yosef Geva (the attaché). He tells me that the ambassador doesn't want to call me directly, but rather that Yosef mediates, in other words call him from my room, and then he will greet me. I have no patience with this stupid game and call the ambassador (Harman)[51] directly. I have known him for many years, so he obviously must do the right thing, and invite me for dinner. During our telephone conversation, he tells me that Abe Feinberg[52] has called and told him that Rostow[53] wants to meet with me. Feinberg asks that I speak to him first before my meeting with Rostow.

I make an appointment with Rostow's secretary for a meeting at the White House tomorrow noon. In the evening, the ambassador invites me to dinner at his home with his family. Most of the dinner is spent arguing—Harman is passionately anti-Rafi.[54] After dinner, we return to the subject of Vietnam. Harman sees great danger in my visit—or to be more precise what I would write about the current military and political situation. 'It's a hornet's nest,' and everything I will write will be bad for the Jews. In any event, his opinions of issues he feels to be sensitive are useful to me, and I will obviously consider them when I do my writing.

I call Abe Feinberg from the ambassador's home. Without further ado, he tells me that President Johnson is being mercilessly criticized because of his policy to both Vietnam, and that Jews—in their different ways—are amongst his harshest critics. Therefore, it would be good if someone like me—who has the ear of the Jewish community—would explain the importance of America's war in Vietnam. He suggests that, after my meeting with Rostow, I should announce that American action in Vietnam is—from Israel's point of view—a test of America's keeping their promise of coming to the aid of nations who are attacked. American Jews would thus understand that the good of Israel (who, according to Feinberg, is one of the nations to whom America has promised aid, if attacked) requires that America fulfils its commitments to aid South Vietnam, which has been attacked by the Viet Cong. I tell Feinberg that I refuse to do this, and that if the meeting has been arranged for this purpose, it would be better to cancel it. I don't want to exploit Rostow, nor waste his time. I tell him that I will make no announcements on Vietnam at this time and that I would only do so after visiting there and seeing and evaluating facts on the ground. Perhaps at the end of my tour I will conclude that there is a connection between Vietnam and Israel—and I will give my opinion of this as well. I cannot say that my words please Feinberg—the ambassador takes no specific stance on any of this.

At the White House and the State Department
17 July 1966
Washington, DC

Noon meeting with Rostow, advisor to the president on national security affairs, in the White House, lasting about 45 minutes. I don't know if this is a response to my request, or his own initiative, as Abe Feinberg had said. Afi said that he is a Jew who plays down his ethnicity. Energetic, clear-headed, looks you straight in the eyes, and acts naturally (puts his feet on the couch, almost in Eastern fashion). His speech is precise, perhaps a little too mechanical. I can't escape a feeling of intransigence, and a calculating approach. But it is entirely likely that it just appears so to me—especially after my having been in Paris and London. The French like flowery speech, while the English speak more modesty—'I don't think that,' etc.

In any event, my main impression of Rostow is his mechanical approach (particularly his forecast of future events). As soon as I enter the room and present my questions. I tell him that I am travelling to Vietnam (he nods —'I know'), and that I have come via France and England. There is confusion about America's objectives (he interrupts—'I am very surprised; the president has made this perfectly clear in his speeches'). I ask:

1. Do you intend to remain in Vietnam permanently, or to leave it? If you intend to leave—under what conditions?
2. Have you agreed to a Communist government in Vietnam—do you choose free and fair elections?
3. Will you agree—after peace has been settled—to unification of the South with the North, if this is the will of the Vietnamese people? I tell him that the query is about long-term policy, 20 years or more, from a strategic point of view.

His answer to the first question, which he repeats and emphasizes, is that America is ready to leave Vietnam 'if and when' peace is assured and after their responsibilities regarding security guarantees to the Southeast Asia Treaty Organization have been fulfilled.

I persist and ask whether—from strategic and political self-interest, not responsibility to others—America sees the need to remain in Asia, and where? (I continuously excuse the harsh, vulgar and primitive nature of my questions. Rostow tells me not to apologize—my questions are good). His reply is that the United States does not want the hegemony of only one power on the Asian mainland.

I ask him if that means—in a positive way—that America will maintain its forces on the Asian mainland no matter what—his answer is yes. (He repeatedly says with a smile that he prefers a negative wording—not one forced on its own. But his meaning is clear—United States global interests do not allow it to relinquish mainland Asia). During our discussion, it becomes clear that his main intention is Thailand. Secondly—he does not feel that America will be requested to leave Vietnam so soon.

I mention U Thant's statement that America is investing a hundred million dollars in an airfield in Vietnam (he interrupts me—much more), and that many countries would regard this as a large sum of money, which would not be invested if they intended an early departure. He smiles:—not tomorrow, only when there will be peace, and only if the Vietnamese government asks us to leave.

He answers the other two questions positively as well. If free and fair elections result in a Communist government, and the South wishes to unite with the North—the United States will not prevent this. But he does not believe that things will develop this way: the Vietnamese people have had enough hardship under communism. Many are refugees who fled Communist rule in the North—For this reason, Ho Chi Minh has lost the heroic halo of national leader in the eyes of the South.

I ask whether he thinks that the coming elections (September of this year) will be free and reflect the will of the people. He answers confidently: yes.

I continue and say he must surely know about U Thant's unwillingness to take UN responsibility for these elections, because he feels that they will not be free. He replies—U Thant has not been there and is not an expert in these matters. I persist, saying that U Thant has told me about the publication of bulletin—number nine?—which disqualifies many voters. He says that that is true. The Communists will not be allowed to vote or be voted for, because they are fighting against the government. When I disagree ('you said that it was possible to shoot them, imprison them—or let them vote'), he repeats that they would not be allowed to vote. I ask what happens to those who want peace, neutrality, or unification with the North. He says that he will let me have material about the terms of the elections. In any event, about 80–90% of the population would be allowed to vote: elections will be free and reflect the true will of the South Vietnamese people.

He doesn't feel that future events would lead to unification with the North. Current developments in Asia are proceeding in a direction favorable to the West. China's leadership role is not accepted in Asia: it has a senseless foreign policy. Its economy—the only one of its kind in the world—has deteriorated during the past 7 years at a time when other countries are advancing. China's leadership has recently suffered costly losses—he mentions the upcoming visit of Ne Win, President of Burma, to the United States. Countries want to unite, not split up. They desire economic development, something only the West can give them. Anyone who has had experience of Communists or their emissaries loathes them. Therefore, when he gives a positive answer to my question 'what will happen if'— he doesn't believe that this would occur, but rather that China itself would be forced to change its policy, limit its birthrate, develop industry, etc. It cannot expand and already sustains only two crops a year (at a time when India, Japan, etc. can potentially overtake her agricultural production, by means of intensive methods which they do not yet have).

Rostow's opinions on anticipated developments are mainly based on economic issues. I mention that nationalism may overcome economic concerns. For example, Jordan got rid of the British, despite giving up economic benefits. Rostow feels that nationalistic motives do not make Vietnam wish to unify with the North or get rid of foreigners (Americans). In any event, at this stage the proximate factor is the war.

I refrain from asking him questions about military tactics. But he, on his own accord, says that the Viet Cong are directed, reinforced, assisted, etc. from the North, beyond the seventeenth parallel. Any decision about stopping the war will come from the Northern leadership. For this reason, the Americans are striking the North to make them recognize that the price of continuing this war is too high, and that it must be stopped. If the North would decide to stop sending reinforcements, Viet Cong forces in the South would disintegrate within 6 months (from the time the North made such a decision).

At the conclusion of the discussion, I tell him that I still want to touch on one subject. (He says that he has 5 minutes, after which time he must meet with the president).—activities of Israel in developing countries. I tell him I want to see Israel as an important center for such activities. The hub should be in Israel, where preparations and planning would occur—then, its people would be sent out to South America, Africa and Asia. However, this must occur conditional to American approval. For example, I am convinced that the Americans were against our activities in Iran.[55]

He replies that he attaches great importance to such activities by Israel, also from a global perspective. There are two nations in the world who receive more aid than anyone else: Israel and Yugoslavia (I think, but am not sure, that he said Yugoslavia).

I tell him that my government opposes the action in Vietnam, but that we could widen our activities in other regions: Thailand, Ceylon, etc. Rostow says that he supports this. He responds positively to my question as to whether Komer—with whom I was about to meet—is the correct person whom to address for this issue.[56]

On the way out, I meet General Taylor[57] (who is apparently accompanying Rostow to the president. He remembers that we are due to meet tomorrow.

My conversation with Rostow is informative, He didn't evade any questions—however I am not convinced that even he believed all his answers to be frank and open—especially about elections in Vietnam this September. I don't believe he thinks that these elections would be free, democratic, or reflect the will of the people.

He surprises me with the clarity of his statement that the United States would not leave mainland Asia. (I asked him whether this was their understanding: to destroy, was it enough to be in outer space and underwater, or everywhere on earth—because they were able to do this with atomic missiles? By contrast, to influence and get involved in local disputes with conventional weapons, nearby bases are required. Rostow agreed completely: that was their approach).

The main thrust of his words are a theory and world-vision of progressive rapprochement of Asian countries with the West based on economic issues. Maybe he is correct: my hesitation stems from my experiences in the Middle East—the Arab countries, where, in the years after the Second World War, nationalism has predominated.

Before we part, he asks that we meet again when I return from Vietnam and promises to read what I wrote attentively. As 'provisions for the road,' he gives me his talk at the War College of June 28, 1961, and that of Lee Kuan Yew, President of Singapore,[58] of June 15, 1966. He promises to send the material on the election to me at the embassy. When we speak of the possibility of America leaving Vietnam, Rostow raises the point that if America was only in Thailand and nowhere else, a large-scale war could develop. Therefore, it was better that American forces be in several countries in the region at the same time.

18 July 1966
Washington, DC

At 10.00 am, a courtesy visit by Ambassador Raymond Hare.[59] He will retire from his position as head of Middle Eastern affairs in a few months' time. We last met in 1949—in Jerusalem, immediately after the fighting.

We have a general conversation about different subjects—the security situation on Israel's borders, agricultural development in Iran, relations between the Iraqis and the United States—a pleasant, non-binding discussion. Hare acts like a man already not involved with his profession.

I am impressed by the quiet in the State Department. Had I just walked through the corridors, I would have thought nobody was there. But on entering a room it is very clear that this is the very nerve center of the Western world.

Meeting the White House at 11.00 am with General Taylor, Special Advisor to the President. The conversation lasts for about 30 minutes, and I am dissatisfied with it—maybe through my own fault. In any event, I cannot move the conversation to anything but formal issues.

Both General Taylor and I know the subject by heart: he does not deviate, even by one word, from his testimony of 17.2.1966 to the Senate Foreign Relations Committee. There are four strategic objectives:

1. Improvement of infantry efficiency;
2. Exploitation of American superiority in the skies over North Vietnam;
3. Maintenance of South Vietnam's economic and political framework;
4. An honorable settlement with the North.

The 'Taylor doctrine' states that the American Air Force has three objectives in bombing the North:

1. Diminish Northern infiltration to the South as much as possible.
2. Convince the South that they are attacking as well as being attacked—and in this way strengthen their confidence in victory.
3. Convince the North that they are paying dearly for the war and make them want it stopped.

After Taylor announces his formulas, I asked him about ground forces. In his opinion they are 'making progress;' they have killed more than 8,000 of the enemy during the past 2 months (about 1,000 a week). I pose two questions:

1. Do reinforcements that arrive not replace the losses? He replies that these numbers are contradictory and again indicates the success inherent in such high enemy casualty rates. I cannot bring him to say whether this is—in fact—the right way to end the war. Despite their losses, the numbers of the enemy are increasing.

2. The second question concerns tactics—what will happen if the Viet Cong change over to fighting in small units? Are the American forces and tactics suitable to such a war against units of eight to ten instead of divisions and battalions?

Taylor gives a partial answer. If indeed the Viet Cong give up on concentration of large forces (one way or the other, to destroy or overwhelm the American Army), their losses will be very heavy. I agree, but remark that the North may well prefer this over defeat and stopping the war. I ask again about American tactics should this occur. Taylor says that this would be a problem, which I should better discuss with commanders on the ground in Vietnam. (Taylor doesn't remember—or want to remember—sections of speeches by Khrushchev, Kosygin[60] and the Chinese that he himself, as well as others, have cited: for a people fighting for national liberation, a guerilla war is the right way to bring about a Communist revolution).

Regarding the air force, I ask him whether bombing defined targets—not necessarily the most sensitive ones—would bring Ho Chi Minh to the bargaining table. (I feel that my question is theoretical, and that I am not suggesting bombing other targets). Taylor says that this is indeed a question. But he highly doubts whether Ho Chi Minh would rush to the negotiating table, even if Hanoi were bombed to rubble.

To sum up their strategic achievements, he says that:

1. Good progress is being made improving activities on the ground.
2. The goal of encouraging the South by bombing the North has been reached.
3. There has been economic and administrative progress, although a crash or retreat could come at any time (he is obviously referring to governmental instability, inflation, etc.).

4. There is no current progress bringing Ho Chi Minh to the conference table. However, he has no doubt that continuous bombings are a heavy burden on the North, who are being forced to look for ways to prevent this.

I ask whether he is worried about the fact that America is waging this war alone and without support, especially of the bombings. What is his reaction to British and Russian recommendations to stop the bombing? Does he think that stopping the bombing will bring Ho Chi Minh to the bargaining table? Taylor says no; as for world public opinion, he believes that NATO supports the United States, despite distancing themselves from certain steps (like England, with bombing of the North). But the main thing is, that the United States is doing the right thing—he has no doubt of that. Regarding criticism, there are differing opinions in the United States as well. Taylor emphasizes these last issues: I think that his words are really meant for General Gavin (whom, as I tell him, I am going to meet on Thursday).[61]

In conclusion, he says that he envies my travelling to Vietnam. I don't believe that this is flowery or forced. Taylor really wants to be in Vietnam—not as a guest from Israel, but as commander of American forces. He exudes military activity. He is tall, erect, and handsome, speaks incisively and looks you straight in the face. He has endless self-confidence.

At the door he wishes me success and says that I should be careful to avoid 'the many difficult bullets.'

America is prosecuting the war in Vietnam alone, although it has partisans amongst its many envoys, and there are political, social and military differences of opinion on how to conduct it. The issue has therefore become the subject of intense and constant discussion, as reflected in Taylor's words. He not only explained military objectives and considerations, but also presented arguments against opposing views, and answered objections which I did not raise. It's almost like we say: 'Someone from *Zichron*

Ya'akov answers a question 'how do you do?' with—'so what's it to you if I'm from *Zichron?*'[62]

The fact that this war is limited doesn't make it any easier. According to the Taylor doctrine on territory (don't cross the seventeenth parallel), war aims (don't bomb cities), types of arms (no nuclear weapons) and political objectives—America must not appear responsible in Vietnam, only as facilitators.

Lunch at 1.00 pm in the state department Rodger Davis, Deputy Assistant to the Secretary of State for Middle East and Asian Affairs.[63]

Davis—a young man—receives me in a most friendly way. About twenty men sit around the table. At the start we discuss Israel and its neighbors, and after that, 'on the other side of the table,' as it were—my upcoming trip to Vietnam. He appears to know my opinions—at least what I expressed ibn my latest article in *Ha'aretz* regarding direct contact with the Arabs. He tackles this issue at once. I explain my stance, and strong objection to the United States on the refugee problem. I see no signs that he accepts my approach. When we start talking about Vietnam, he—like all the Americans I met—maintains that the entire world not only allows America to shed its blood, but derives benefits from it, while acting self-righteously.

At the end of the meal, he rises and proposes a toast in my honor and blesses my journey. I use the opportunity to address a few words of thanks for the assistance and hospitality that I have received. The atmosphere of the meal is very pleasant. Colonel Hayes remarks on it as well afterwards.

Meeting at 3.00 pm with William Bundy (brother of the 'great'. McGeorge Bundy, past presidential national security advisor and now President of the Ford Foundation).[64] No new issues are brought up. Bundy is an intellectual with a talent for speaking (which he enjoys). For some reason, he feels that the situation in Vietnam has much in common with that in Greece at the end of the Second World War.[65] When I ask why India disagrees with America's bombing policy, he replies that she (in other words Mrs. Gandhi[66]) has to 'take the Soviet Union into consideration.' India wants the Soviet Union to supply them with arms, because problems with Pakistan don't allow the United States to fulfil their needs. Apart from that, India knows that America will not stop economic aid, even if it disagrees with their bombing policy.

I ask which countries in Asia unreservedly support the United States in its Vietnam policy. He enumerates—Japan, Thailand, Malaya, Philippines and Singapore. Indonesia also has expressed support in private discussions but cannot suddenly make a sudden 90° turn away from Sukarno's previous declarations.[67] Up to now, Burma has not said anything against the bombings, and Ne Win is about to visit the United States.

Of course, America has an explanation for every nation that is against the bombings. Their explanations may be invalid, but there is something in their complaint of national double-standards—the many statesmen who, privately, say the opposite of their public pronouncements. My final state department visit today is with Director of Information Chief Leonard Marks.[68] He is interested in the goals of my trip and promises the help of his staff in Saigon, etc. The conversation goes well ... until I make the mistake of saying that I don't believe America is properly explaining (in other words failing to explain) its international concerns. (He agrees that world opinion is against them, but in his opinion the problem lies with them, not the United States). I bring up the issue of elections, saying that I, for example, am not prepared to 'buy into' the upcoming September elections as being free and fair. I could be persuaded that completely free and fair elections cannot be held in time of war. However, when they aggressively demand that the U.N. and Western public opinion accept these elections as free, when they invalidate Communist and neutrality candidates (who demand that Vietnam be a neutral country)—they are hypocrites. Marks seethes with indignation, saying that the definition of 'free' is a semantic one, which everyone can define as they wish. Foreign Secretary Rusk[69] compares entry of the Communists into the government—if they are elected—to entry of a fox in a henhouse. The conversation deviates from plan, but this does not upset me a bit and I say to Marks: 'you asked me for an example, and I responded. But the reasons you have given why I should accept the elections as free and fair only serve to emphasize the questions I raised. I am not here to judge the fairness of your actions. But I am not prepared to accept incorrect definitions, just because you consider them suitable in wartime.'

My impression until now has been that the Americans are not prepared to listen to opinions of their peers. Perhaps this is caused by the sharp internal arguments they are having about Vietnam. They have no patience for foreigners who come just to observe and criticize—without taking any responsibility.

Dinner at 8.00 pm with Joe Alsop. We are about twenty—journalists and statesmen; among these Secretary of the Treasury Fowler and Senator Symington (with whom I had breakfasted with Teddy Kollek, when he was in Israel).[70]

Alsop is aggressively optimistic about Vietnam. The war will be resolved by November, and next spring (April), will be the beginning of the end. No self-respecting journalist has not visited Vietnam several times. All are full of enthusiasm for the fighting spirit of units they have visited, and the technical and logistical means the army has developed to fight a guerilla war in the jungle. However, their enthusiasm for military prowess of American units lacks analytical investigation of the not inconsiderable number of failures (not defeats, but failures) that have been 'missed' due to faulty information or concepts.

Much criticism has been levelled at other (Western) countries who are not really interested, but only criticize, the American war in Vietnam. But the Americans admit that Vietnam will not be the decisive factor in the upcoming American elections: rather, local issues, racial disturbances, delay in achieving the aims of the 'Great Society,' will predominate. Strange; Americans want others to see Vietnam as the central issue, but they don't give it first place in their own considerations.

19 July 1966
Washington, DC

Today—discussions about American civilian developmental activities in Vietnam. Morning in the state department. Meeting with Rutherford Poats,[71] deputy director of the agency for international development. His area of expertise is East Asia.

We sit in a specially arranged room, his assistants and their clerks around a long table. Each one in turn stands and explains the structure and organization, and their specific roles in it, with a blackboard and diagrams hung on the walls. They are trying to achieve the following:

1. Stabilize the economy and prevent inflation (by means of selling supplies and equipment sent from the United States, using Vietnamese currency).
2. Agricultural development: seeds, irrigation, chemical fertilizer, insect repellant, etc.
3. Services: paving roads, building hospitals, schools, etc.
4. War against Viet Cong terror. For this purpose, the country is divided into provinces, then into regions and villages. Each village has an 'elder,' and each province an American representative, whose job it is to provide aid.

The people give the impression of knowing their subject well: they are young and decisive. But the feelings of hesitation that I feel during the lectures and explanations will not go away. Does reality in Vietnam really correspond to this 'black and white,' decisive approach? Perhaps the Vietnamese psychology is different? In the latter case, all this organization will be of no use at all. In any event, I will only be able to see this when I am there—perhaps not even then.

On my request, they also explain their regional activity—the Bank of Asia—which is being made ready. I say that it is in Israel's interests to participate more actively in Asian developmental projects (outside of Vietnam for the moment, as per our country's policy). We arrange that the American Embassy in Saigon will set up a meeting for me with the relevant people in Bangkok, where I would have to fly from Saigon.

Two main problems stand out during these morning conversations:

1. Concentration of the population on the coast, leaving the mountainous region almost unpopulated and easy for the Viet Cong to seize, and profit from—by establishing mountain bases, from which to attack the coastal population. I ask whether there are any American plans to establish settlements in the mountains. They answer no: the mountain tribes are half-wild: they sharpen their teeth like those of wild animals and are naked except for loincloths. These areas cannot be developed for agriculture therefore activity must be limited to the plain and coastal areas. No matter the explanation, this division of the agricultural community makes security issues much more difficult and complex.

2. In answer to my question, I am told that acts of terror against those who fill positions of responsibility in South Vietnam have not diminished. If this is so, I have no idea what the meaning of the 50% and 60% which the Americans say they control, means. Does everyone who takes on a government position sign his own death warrant, with the South Vietnamese government and military powerless to stop it? This surely means that there is no foundation for establishing an administration capable of implementing this development.

Lunch at the Metropolitan Club with Eric Sevareid, journalist and television reporter.[72] He is against the war—and doubts the wisdom of how it is being conducted. The planned framework is not being adhered to—indeed, it is being deviated from and broken on all sides—and he sees no end to the conflict. The United States is not suffering from the war economically or financially, but it is also true that many necessary developmental projects related to the 'Great Society' have had to be postponed because of the war. There is a limit to peoples' energy and attention span—including those of the president and government. While they are so preoccupied with Vietnam—nothing else is dealt with.

Afternoon meeting with Robert Komer, Special Assistant to the President. Although he has visited Israel several times in his previous capacity (senior member of staff for Middle Eastern Affairs at the National Security Council), this is our first meeting. For some reason, he starts with a lecture to me about our exaggerated suspicion of the Arabs; time after time, we announce loudly that the Arabs are about to attack us, but this doesn't happen. I explain my position to him, rejecting American guarantees. I tell him America should deal less with problems between us and our neighbors, and more with our cooperation in developing countries. We return to the subject of Vietnam (economic and agricultural development, and security in the villages in the face of Viet Cong terror). He emphasizes, in no uncertain terms, how important he regards it for Israel (personified by me) to support the war in Vietnam. This is a repetition of what Abe Feinberg had said (only after he starts to speak do I understand why he asked that I come without the colonel who acted as my escort). I tell him that I am on my way to, not from, Vietnam. After I have been there and formed an opinion—I will not hesitate to share it, no matter what it will be.

While leaving, he asks that I convey his greetings to Shimon Peres and tell him that he is in agreement with his approach to security, but in terms of foreign policy (towards the Palestinians)—he is a fox. I promise to do so.

McNamara

In the evening—dinner with the publisher of the *Washington Post* and Newsweek—Katharine Graham.[73] The surprise of the evening, for me, was the presence of the secretary of defense, Bob McNamara.

In contrast to his photographic and television image,—he is pleasant and converses easily, without any signs of inflexibility. In contrast to the previous evening, the number of guests is small. Apart from two—a journalist from the *Daily Telegraph* and one from *Newsweek*—whom I had met the previous evening—there are also several members of the press, W. Rostow (whom I had met the previous Saturday), the secretary for education, McNamara and his wife. (His wife told me, amongst other things, that their younger daughter is happy to be home, saying that her father said it was permitted to work on the [Christian] Sabbath.)

Naturally, most of the evening conversation revolves around Vietnam. McNamara does not avoid the issue (I suspect that he would have been sick of it by now). After I have raised several points, he asks whether I was going to meet with General Westmoreland. I tell him that no meeting had been arranged: he takes out a notebook, writes something in it, and says that he will telegram Westmoreland about my visit: he wants us to meet and talk. He also asks me to visit, and give him my impressions, on the way back. My conversation with McNamara fails to shed light on several issues, which remain vague. Firstly—why don't they allow the Viet Cong to attack during the monsoon season—(in my opinion) that is the time to overwhelm them, in great strength—especially with superior firepower. After all, the Viet Cong have no air support or artillery. McNamara explains again that they have achieved the same thing by previous attacks on Viet Cong bases, but I see that he doesn't really believe this.

The second and most important question, is—security for Vietnamese civilians. McNamara confirms what Komer had said: that attacks on official representatives of the South Vietnamese regime have not diminished at all. I ask him, if that was the case, what is the meaning of being in charge of 'half of the region?' He agrees that these numbers don't mean much and repeats his assertion to those responsible 'that they have not even achieved the first square mile of pacification.' (He adds that he is probably exaggerating, but so far he sees no real results in this most important of areas—civilian security). He recognizes how important this is, and how little has been achieved so far. He emphasizes, and repeats several times, that this is the task of the South Vietnamese leadership, and not within the capability of the American military.

I don't know whether the Americans can really provide security for the population against Viet Cong terror attacks (this depends on familiarity with the area, long-term perseverance and presence of many soldiers—the Americans' aim is a 140,000- strong Vietnamese police force). To expect that, under the present conditions, the Vietnamese Army can do this—sounds like impossible optimism. I am not the only one in the room who isn't clear what McNamara really means.

Viet Cong terror attacks against government officials have three important intents:

1. Condemnation of collaborators as traitors to the homeland (cooperating with the American enemy): these are put on trial and summarily executed;
2. Lack of upper-level staff, preventing the South Vietnamese from establishing a stable administration and government. The result—civilian anarchy;
3. 'Educational' influence. This is something similar to what Taylor testified to Congress: The American aim is not to destroy the enemy, but to 'help them see the light'—to change their direction.

There is, to a large extent, a parallel between Viet Cong terror attacks and American bombing of the North. The Viet Cong want to convince the Americans that there is a 'high price to pay' for their being there. Terror tells the South Vietnamese: get rid of the Americans, and there will be peace.

The vaguest issue of all is the question of ending the war, or progress towards victory. McNamara says that the North's losses in the South (1,000 a week) will be of crucial importance. However, I doubt the veracity of the number. I remind him that only 10% of the North Vietnamese are currently in the South. He himself knows this (and agrees), but this comes as an unpleasant surprise for the others.

In conventional wars, clear indication of progress towards victory is geographical and territorial. The army must reach Berlin, Paris, etc., to conquer the enemy's capital, make its government submit, and sign a surrender agreement—with the new government. But here—from the beginning, the Americans have announced that they will not cross the seventeenth parallel, nor bomb civilian targets. How can the matter be settled? How can the enemy be forced to surrender? How can progress be measured? What is the true measure of progress?

Quite obviously, I am not the first to have posed these questions. I am pleased that McNamara does not evade but is willing to revisit and discuss them. But I am sorry not to receive a reassuring answer. There are also no clear answers to the goals and objectives of this war: if the question is posed, Americans simmer with anger. They are sick of repeating themselves, saying the same thing over and over. Their goals are purely democratic: restraint of aggression, allowing a weak ally to live as it pleases, without being invaded by its neighbors who force their will on them. However, the more simplified the explanation, the less clear it becomes. If this is the case, why don't countries who aspire to freedom and independence—like the countries of Western Europe, and progressive elements within the United States—support this war? If the argument is so clear, why hasn't the enlightened, progressive world not been seized with enthusiasm and positive opinions of American military activity in Vietnam?

Someone (I don't remember who) around the table gives the accepted answer, apparently widespread in American military circles: The French (and British) are simply envious of the Americans, who succeed in places where they fail.

To properly define reasons for negative French and the British attitudes, it seems more accurate to say that both nations have had bitter experiences of imperialism, lessons from which they have learned. They already have 'developed' 'far-away,' backward countries and 'defended' them from neighboring aggression. But, at the end of the day, they have concluded that they must leave and let each nation come together in their own country. France and Britain see the United States as repeating their own past mistakes. They regard American action as a kind of late twentieth century imperialism—something with which they fundamentally disagree.

McNamara indicates that the war in Vietnam does not burden the American economy. Obviously, it costs taxpayers' money, but it doesn't delay American economic development. Others have been proud to agree with and support his point of view. Whatever the opinion of the war, it is impossible not to mention the activity and enormous growth potential of the American economy. Even he is only occupied with Vietnam for about 40% of the time—some of his government ministers much less than that.

Rostow is very friendly: amongst other things, he suggests that I stop over in some place or other (I don't remember where) and write my report. This is what he does when he reports on Vietnam with General Taylor. I thank him and say that I do not intend to write a formal report. I do not have a deadline for my summary, which I will write in Israel. He also asks me to contact him and give him my impressions if I return via Washington.

McNamara requests details of how we train the uneducated in the Israel Defense Forces. He is very enthusiastic and wants Congress to approve such a measure. (At the present time, those with lack of a minimal education are not admitted into the American Army). This is principally necessary for blacks lagging in education: he said that the South would oppose such a measure.

When we part, he says again that he will write to Commanding General Westmoreland about my visit and asks me to contact him on my way back, to give him my impressions.

Outside Experts
20 July 1966
Washington, DC

At 10.00 am, my lecture at the Institute for Defense Analyses on 'The background of the Sinai Campaign and comments on its outcome.'

Lunch at the *Bank Le'umi* with David Kochav[74] and some of the directors.

Dinner: reception with Israelis.

Tomorrow, on to Boston.

21 July1966
Boston

At noon: Lunch meeting with General Gavin.

He is of the opinion that Westmoreland is not sufficiently strong-willed to be a suitable commander in Vietnam. This, in answer to my question why the Americans have prevented the Viet Cong from attempting a monsoon offensive. He is also critical of McNamara—he is not qualified to conduct this war or oversee units and is trying to make decisions (on how the war is conducted) that should best be left to military professionals.

I don't ask him many questions about Vietnam, because we are not alone, but he confirms that his opinions are the same as those expressed in an article that he sent me.

Before we go in for lunch, his [female] secretary tells me that our Embassy is looking for me. I call them and am told that McNamara wants to speak to me. I telephone him; he is at a lunch meeting, but he comes at once to the telephone and tells me that he wants to read me a telegram received from Westmoreland in answer to his own communication. Westmoreland replies that 'he would be pleased to meet with General Dayan and assist him in his visit to Vietnam.' I thank McNamara very much—actually, I am amazed, because this is no ordinary gesture (I have no idea why I merit it): he could have told his secretary to tell our military attaché that he has approved my trip. I tell McNamara again that I will be in touch if I return via Washington—again he requests that I do this. (During the evening, I find out that the fact that McNamara was looking for me has come to the attention of our defense department delegation in New York. I will 'never hear the end' of this).

Brown brings the dinner in my honor forward from 6.00 pm to 5.00 pm, to give us more time—the flight to New York leaves at 8.00 pm.

Apart from Brown, Carl Kaysen, Lucian Pye, Milton Katz, and (Matti) Peled are dinner guests.[75]

We begin talking about the Middle East and then turn to the two main issues. Firstly, I ask whether there is room for cooperation between Israel and the United States in third world country development—especially Latin America and Asia. After finding out what we propose to do, they reply that there is both place and use for this, on condition that 1) It is done under international auspices. The United States would not accept a two-nation agreement; 2) We would be patient with all the bureaucratic obstacles and difficulties inherent in any cooperation with the United States government—this struggle was 'enough to drive anyone crazy.'

They ask me why I am interested in such things. I tell them that it is good for Israel for three reasons:

1. Israel must join in constructive international affairs, not remain mired in problems with the Arabs.

2. We currently have a surplus of suitable manpower.

3. We must 'snatch victory from defeat.' We started this 10 years ago and failed; we learned from our mistakes, and now it is 'working.'—It is worthwhile to widen our efforts. However, the United States is the chief financier of third world aid, and without its support Israel has no chance.

After that, I move on to the second and principal subject—Vietnam. During the conversation (which lasts about an hour), I ask three questions:

1. How do the Americans intend to end the war?
2. What are the United States long-term strategic objectives in the region?
3. Why—if American objectives are not merely self-interest—is there no support for the war from Western Europe?

Concerning the first question, I say that Britain and France (without generalization) believe that The United States must:

1. Not fail;
2. Not leave immediately, because this would be interpreted as a retreat;
3. Terminate the war—by stopping the fighting.

The war is currently at the initiative of the United States. If they stop their attacks (bombing the North and attacking in the South), the Viet Cong will surely not start new attacks. At most—they would increase infiltration into the South but maintain their forces in place. At the end of the day, the South and North will unify under a Communist government. Those to whom I spoke in Britain and France are prepared for this. What alternative does America suggest or plan? The main responses come from Kaysen and Pye (although others try to interrupt). America sees two possibilities: first, the Viet Cong will be broken, and come to the bargaining table, to reach an agreement with the Americans; second—they will get tired of fighting, and the war will gradually fade away. An agreement with America would be advantageous to the North, because then America would be prepared to leave Vietnam. The president has said that this could act as an enticement to force the North to the bargaining table which would allow them to 'save face' in exchange for a signed agreement with the Americans.

In the second case, no agreement will be reached but the North will simply stop sending reinforcements, and the Viet Cong in the South will be killed, scatter, and retreat to the North.

The same actions are needed to achieve either of the two objectives: the Americans must continue to bomb the North, and search-and-destroy the Viet Cong in the South. Pye believes that, if the war stopped now, there would be no more infiltration from the North. It seems to me that Pye is of the opinion that—quite simply—hostilities should cease, even though he doesn't spell that out.

Things become clearer when we get to the question of America's long-term foreign policy. Before we start, those present emphasize that they express the official American stance, not their private perspectives (Katz says—simply—this entire war is crazy, and it would have been better if we had not entered it. However, whatever happens in Vietnam, after we have already gone in—even he doesn't suggest that America withdraw and leave Vietnam now).

They explain that, in the long term, America wishes to retain a presence in South-East Asia—I believe that this is their opinion as well. This, to draw a line and put a stop to expansion or Russia, China, etc. America must have a presence in the area, but not necessarily in Vietnam.

The conclusion of the discussion is the most piquant—after they have heatedly explained the need for the war—they severely criticize Europe because, in fact, they have been the recipients of American largesse (all these years), without properly sharing the burdens—Kaysen emphasizes this point. However, the sum and substance of everything is that America has become entangled in Vietnam: if someone would take Johnson into the corner of a room and show him a way to leave Vietnam without destroying American prestige (president, cabinet, army commanders, etc.), he would thank him very much and withdraw the army ...

When we part, I ask Brown that if they have a debate on Middle Eastern affairs there in the future, they should let me know, and I would try to attend.

The dinner conversation with Brown's group is the best I could have wished for, before I summarize my 'Washington seminar.' Definitions are clear and exact, and I am encouraged by the warmth and interest of those taking part. Apart from Brown, they all participate, repeatedly interrupting each other. Like others, they become very indignant when I mention Europe's lack of support. At the same time, they hotly contend that, if they would say this in front of the Europeans, it would only alienate them. I don't know whether they realize how out of proportion, though understandable, their sensitivity is to domestic public opinion as compared to their disdain for the Europeans, whose views will not change anything and who will always have to be bailed out by America. In any event, it will not change their opinions (or actions). America will always come to Europe's aid, as it did in both World Wars, Korea, etc. It is understandable if they don't take Europe's stance into account from a practical point of view; however, if the war in Vietnam really is a war of Democracy against Communist aggression, then the Free World should applaud, and appreciate what they are doing: instead, all they get is criticism. This offends them, although (it appears to me that) they see that criticism as justified. At the end of the day, everyone present agrees that—even if the Americans achieve their goals—the net result would be (they regret to say) America's having to maintain an army of occupation in South-East Asia.—Formosa (navy), Korea, Thailand, Vietnam, and other locations.

Katz is the only one to say firmly that he sees no crucial reason for the United States to be in Vietnam (or the surrounding region) for the foreseeable future.

Everyone also agrees that the present situation has been brought about by a complex chain of events and not caused on purpose.

I ask how Russia and its Communist allies would react to the 'fading away' of the war, which the Americans want to achieve, even if it is one-sided for a long time. Kaysen maintains that the Russians' quarrel with the Chinese has diminished. The Chinese have prevented the Russians from using their railways, demanding that they send provisions and supplies to the Viet Cong by sea. The Chinese have done this to bring about confrontation between Russia and the United States (on the high seas), requiring that this time the Russians not surrender to the Americans and run away, as they did in Cuba (during the missile crisis). Now, the Russians can tell the Chinese that they cannot use the sea route, because the Americans are bombing the harbors.

One other point regarding the Russians: their aid (as is the case with China) is limited to support of the North. They don't want it to fall, so they send missiles, planes, etc.—just for the North Vietnamese struggle. (For this reason, those present believe that there is no reason to suppose that the North would increase infiltration of the South; their objective is only to protect the North). They don't think that the Soviet Union will make the war worse.

I look into the possibility of travelling to Burma and am told that nobody knows why Ne Win wants to visit the United States. They regard relations with Burma as important, because it is more independent than other countries in the region.

In the meanwhile, I am completely exhausted. My flight to New York is delayed for 3 h. Last night, I didn't sleep at all. I think that the first thing I'll do when I arrive in Vietnam—is to find a nice, leafy jungle tree, and fall asleep under it.

Honolulu-Tokyo (by plane)

Americans are making a mistake ignoring what the Europeans think: they don't distinguish between governments and public opinion. Even if American complaints against governments are justified—because they don't pull their weight in a specific struggle—they cannot hide from public opinion. This is the opinion of the clergy, diplomats and others—who do not bear responsibility for their government's foreign policy—and are free to express their opinions, as they wish. They must deal with the crux of issues—by argument. But Americans, instead of taking counsel with them on issues such as America's place in Asia, reject their opinions, because they regard their country's foreign policy as incorrect. For this reason, they simply do not respond. The absurdity of this approach increases when one considers the fact that, in every instance—no matter how small—American public opinion decides the actions of the government. The fact that 75% of Americans condone the bombing of the North is ironic—as many have pointed out, but there is grave doubt whether this can be determined by surveys. Public opinion is important in issues that affect the people, who must pay the price (by, for example, full military conscription for every 18-year-old male). A 'pragmatic' approach leads to differences of opinions, depending on who is asked what question, and who thinks what is important.

If Westmoreland were to ask me—directly or through his representatives—what I want to see, hear and do in Vietnam, I would provide the following list:

1. Specific weapons classes developed by Americans for guerilla warfare and how they work—both technical and tactical;
2. The tactical principles of different operations (artillery, helicopters, commandos, etc.);
3. Principal operations up to the present time (during the past year or two): what were the expected results, and what really happened;
4. Attachment to an American fighting unit, fighting the Viet Cong on the front line in the near future. (I must be allowed into the most forward echelons: company, platoon, etc.);

5. Attachment to a non-American unit—South Vietnamese or Korean—expected to be active in the near future;[76]
6. Witness 'pacification' operations in different places;
7. Visit (for several days) a village in an area not fully controlled by the South (amongst the 25% that change hands all the time);
8. Use of a translator when I am with the units, or in areas with the locals;
9. An opportunity to talk with those in charge of policy and military activity—Westmoreland's chief of staff, deputy assistant chief of staff for operations or Westmoreland himself.

Endnotes

1 Jean-Louis Nicot (1911–2004), French air force general and commander of the French air transport fleet during the first Indochinese War. Later imprisoned because of his role in the Algiers putsch, but amnestied.

2 The Fourth French Republic.

3 Điện Biên Phủ (13 March–7 May 1954), decisive battle between the French and Viet Minh forces which led to the French withdrawal from Indochina.

4 Exactly what Ho Chi Minh (1890–1969) would have done had the French and subsequently Americans not intervened militarily in Vietnam is an open question.

5 Dayan has a low opinion of the shallow, superficial Californian way of life. He wrongly generalizes it for the entire United States.

6 Mordechai Limon (1924–2009), fourth commander of the Israeli Navy (1950–1954).

7 Marie Joseph Pierre François Kœnig or Koenig (1898–1970) was a French general during World War II during which he commanded a Free French Brigade at the Battle of Bir Hakeim in North Africa in 1942. He started a political career after the war and was posthumously elevated to the dignity of Marshal of France in 1984.

8 Georges Jean Raymond Pompidou (1911–1974), French politician who served as President of France from 1969 until his death in 1974. He had previously served as Prime Minister of France from 1962 until 1968 under President de Gaulle.

9 Pierre Marie Gallois (1911–2010), French air force brigadier general and geopolitician instrumental in the constitution of the French nuclear arsenal, considered one of the fathers of the French atom bomb.

10 Robert S. McNamara (1916–2009), American businessman and government official who served as the eighth United States secretary of defense from 1961 to 1968 under presidents John F. Kennedy and Lyndon B. Johnson at the height of the Cold War.

11 Christian Marie Ferdinand de la Croix de Castries (1902–1991), French commander at the Battle of Diem Bien Phu in 1954.

12 Probably Marcel Dassault, born Marce; Ferdinand Bloch (1892–1986), French engineer and industrialist who spent his career in aircraft manufacturing.

13 Trần Văn Hữu (1896–1984) served as president of the government of Cochinchina during 1948–1949 then as Prime Minister of the State of Vietnam from 1950 to 1952.

14 Walter Eytan (1910–2001), Israeli diplomat who served as director general of the Israeli foreign ministry in 1948–1959, and Israeli ambassador to France 1959–1970.

15 Yael Varad worked in the Israeli Embassy in Washington, DC, contributing to diplomatic efforts and relations between the United States and Israel; David Sheinborn is known for his meetings with Moshe Dayan in 1966. Sheinborn was a notable figure in the field of security and intelligence. His interaction with Dayan, who was the Israeli Defense Minister at the time, was part of his broader involvement in geopolitical and military affairs.

16 General Hubert de Seguins Pazzis (1913–1994) French officer served formerly served under General Leclercq in Indochina (1945–1946) and later under General de Castries at Dien Bien Phu and in Algeria. I thank Georges Salmaso for this information. The reader may imagine the difficulty of transliterating such a name from Hebrew without vowels.

17 Philippe François Marie Leclerc de Hauteclocque (1902–1947), Free French general during the Second World War and, after the war, commander of the French Far East Expeditionary Force. Posthumously elevated to the dignity of a Marshal of France.

18 Arthur Buchwald (1925–2007), American humorist best known for his column in the *Washington Post*.

19 Raoul Levy (1922–1966), French film producer, writer and director; H.J. Kaplan, With the American Press in Vietnam, May, 1982. https://www.commentary.org/articles/h-kaplan-2/with-the-american-press-in-vietnam/. Accessed September 2024; Élie Robert de Rothschild (1917–2007), guardian of the French branch of the Rothchild banking dynasty; Irwin Shaw (1913–1984), American playwright, screenwriter, novelist, and short story writer.

20 Defying standard military practice, General Võ Nguyên Giáp (1911–2013), commander of the Viet Minh Army, brought in vast amounts of heavy artillery (including heavy anti-aircraft guns) and managed to move these bulky weapons through difficult terrain by individual men and women up the rear slopes of the mountains. His twenty-four 105 mm howitzers placed on the forward slopes of the hills around Dien Bien Phu, in deep, mostly hand-dug emplacements protected them from French aircraft and counter-battery fire.

21 Bernard Law Montgomery, 1st Viscount Montgomery of Alamein (1887–1976), British field marshal and one of the most important military commanders of World War II.

22 Aharon Remez (1919–1994), Israeli civil servant, politician and diplomat, and the second commander of the Israeli Air Force. From 1965–1970, he was Israeli ambassador to the United Kingdom.

23 Josip Broz Tito (1892–1980), Yugoslav communist revolutionary and politician who served in various positions of national leadership from 1943 until his death in 1980.

24 After the terrible bloodshed and destruction of the Korean War, this is very difficult to believe but is translated directly from the original. The South Korean contribution to the American effort in Vietnam shows just how little they were interested in Communist-dominated unification.

25 The end of the cold war underlines the soundness of Eisenhower's approach.

26 James Harold Wilson, Baron Wilson of Rievaulx (1916–1995), British statesman and Labour Party politician who twice served as Prime Minister of the United Kingdom from 1964 to 1970 and again from 1974 to 1976.

27 This is likely a reference to Montgomery's *A History of Warfare,* first published in 1968.

28 Sir Anthony Eden, Lord Avon (1897–1977), Prime Minister of the United Kingdom and Leader of the Conservative Party from 1955 until his resignation in 1957.

29 Gamal Abdel Nasser (1918–1970), Egyptian military officer and politician who served as the second president of Egypt from 1954 until his death in 1970. For details of the Suez crisis, see https://history.state.gov/milestones/1953–1960/suez. Accessed September 1, 2024.

30 Ernest Bevin (1881–1951), British statesman, trade union leader and Labour Party politician.

31 The representative body of Jewish residents in Mandated Palestine prior to the establishment of the State of Israel in 1948.

32 Major General Orde Charles Wingate (1903–1944) was a senior British Army officer known for his creation of the Chindit deep-penetration missions in Japanese-held territory during the Burma Campaign of the Second World War. A

dedicated Christian Zionist, he set up a joint British–Jewish counter-insurgency unit called the Special Night Squads in Mandatory Palestine before the war. Killed in an aircraft accident 24 March, 1944; Field Marshal First Viscount William Slim (1891–1970) who led the Fourteenth Army to victory in Burma during the Second World War.

33 General Sir Charles Keightley (1901–1974), Commander in Chief of Operation Musketeer (Anglo-French plan for invasion of the Suez Canal Zone to capture the Suez Canal during the Suez Canal crisis of 1956).

34 The most famous example of Horatio Lord Nelson's (1758–1805) bent to selectively disregard the commands of his seniors occurred during the Battle of Copenhagen in 1801. Upon receiving the signal to break off action, Nelson turned to Captain Thomas Foley of his flagship, Elephant, 'You know, Foley, I have one blind eye and have a right to be blind sometimes. I really do not see the signal.'

35 David Ben Gurion (1886–1973), first Prime Minister of Israel.

36 *Irgun Zvai Le'umi (Etzel)*, Zionist paramilitary organization that operated in Mandatory Palestine between 1931 and 1948 specializing in using terror tactics to drive out the British.

37 Henry Baron Walston (1912–1991), British farmer, agricultural researcher and politician, firstly for the Liberal Party, then for Labour, and for the Social Democratic Party.

38 Zvi Zafriri (b. 1930), former CEO of Israel Railways. At the time of Dayan's visit to London he worked in the Israeli defense delegation to the United Kingdom.

39 Julian Lord Amery (1919–1996), British Conservative Party politician, who served as Member of Parliament for 39 of the 42 years between 1950 and 1992. Son of Leo Amery (1873–1955).

40 Ngô Đình Diệm (1901–1963), South Vietnamese politician who was the final prime minister of the State of Vietnam (1954–1955) and later the first President of South Vietnam from 1955 until his capture and assassination by a CIA-backed 1963 South Vietnamese coup.

41 J. William Fulbright (1905–1995), United States politician, academic, and statesman. Well known (as was Robert Kennedy) for his opposition to the Vietnam War.

42 Shimon Peres (1923–2015), Israeli politician and statesman who served as the eighth prime minister of Israel from 1984 to 1986 and from 1995 to 1996 and as the ninth president of Israel from 2007 to 2014. He was a member of

twelve cabinets and represented five political parties in a political career spanning 70 years.

43 Anthony First Viscount Head (1906–1983), British soldier, conservative politician, and diplomat. High Commissioner to Nigeria from 1960 60 1963 and to Malaysia from 1963–1966.

44 U Thant (1909–1974), third secretary general of the United Nations, Burmese born.

45 Michael S. Comay (1908–1987), Israeli Ambassador to Canada from 1953 to 1957, permanent representative of Israel to the United Nations from 1960 to 1967, and ambassador to the United Kingdom from 1970 to 1973.

46 U Nu (1907–1995), prominent Burmese statesman and the first Prime Minister of Union of Burma.

47 Nguyễn Cao Kỳ (1930–2011) was a South Vietnamese military officer and politician who served as the chief of the Republic of Vietnam Air Force in the 1960s, before leading the nation as the prime minister of South Vietnam in a military junta from 1965 to 1967

48 Ne Win, born Shu Maung (1911–2002), Burmese military officer and politician who served as Prime Minister of Burma from 1958 to 1960 and 1962 to 1974, and also President of Burma from 1962 to 1981. Ne Win was Burma's military dictator during the Socialist Burma period of 1962 to 1988.

49 Major General Yosef Geva (1924–2014).

50 General William C. Westmoreland (1914–2005), commander of United States forces during the Vietnam War from 1964 to 1968. He served as Chief of Staff of the United States Army from 1968 to 1972.

51 Avraham Harman (1914–1992), Israeli ambassador to the United States from 1959 to 1968.

52 Abraham Feinberg (1908–1998), United States, businessman, supporter of Israel and friend of Harry S. Truman.

53 Walt Whitman Rostow (1916–2003), American economist, professor, and political theorist who served as national security advisor to President Lyndon B. Johnson.

54 Rafi, an acronym for *Reshimat Poalei Yisrael* (lit. Israeli Workers List) was a center-left political party in Israel, founded by former Prime Minister, David Ben-Gurion in 1965. In 1968 it was one of three parties that merged to form the Israeli Labor Party.

55 Israel had good relations with the Shah of Iran.

56 Robert W. Komer (1922–2000), American national security advisor known for managing Civil Operations and Revolutionary Development Support during the Vietnam War.

57 Maxwell D. Taylor, senior United States army officer and diplomat of the mid-twentieth century.

58 Lee Kuan Yew (1923–2015), Singaporean statesman and lawyer who served as the first Prime Minister of Singapore from 1959 to 1990, and Secretary-General of the People's Action Party from 1954 to 1992. He was a Member of Parliament from 1955 until his death in 2015. Lee is widely recognised as the founding father of the modern Singaporean state.

59 Raymond A. Hare (1901–1994), United States diplomat who was Director General of the United States Foreign Service from 1954 to 1956 and Assistant Secretary of State for Near Eastern and South Asian Affairs from 1965 To 1966.

60 Nikita Sergeyevich Krushschev (1894–1971), First Secretary of the Communist Party of the Soviet Union from 1953–1964 and Chairman of the Council of Ministers from 1958 to 1964; Alexei Nikolayevich Kosygin (1904–1980), Premier of the Soviet Union between 1964 and 1980.

61 James M. Gavin (1907–1990), senior United States officer with the rank of lieutenant general during the Second World War who latterly became an opponent of the Vietnam War.

62 Untranslatable. *Zichron Ya'akov* is a town in Israel 35 kilometers south of Haifa.

63 Rodger P. Davis (1921–1974), American diplomat, assassinated in Nicosia.

64 William P. Bundy (1917–2000), American attorney and CIA analyst who served as an advisor to foreign affairs advisor to both presidents John F. Kennedy and Lyndon B. Johnson. He had key roles in planning the Vietnam War, serving as deputy to Paul Nitze at Defense under Kennedy and as Assistant Secretary of State for East Asian and Pacific affairs under Johnson; McGeorge Bundy (1919–1996), United States National Security Advisor to presidents Kennedy and Johnson from 1916–1966, and one of the chief architects of the escalation of the Vietnam War.

65 A civil war between Monarchists and Communists resulted in a Monarchist victory.

66 Indira Gandhi (1917–1984), prime minister of India from 1966 to 1977 and again from 1980 until her assassination in 1984.

67 Sukarno (1901–1970), First president of Indonesia serving between 1945 and 1967.

68 Leonard H. Marks (1916–2006), Director of the United States Information Agency.

69 Dean Rusk (1909–1994), United States Secretary of State from 1961 to 1969 under the Kennedy and Johnson administrations.

70 Joseph W. Alsop (1910–1989), American journalist and syndicated newspaper columnist from the 1930s through the 1970s; Henry H. Fowler (1908–2000), Secretary of the Treasury under President Johnson; Stuart Symingon, United States senator from Missouri from 1953 to 1976; Theodore (Teddy) Kollek (1911–2007), mayor of Jerusalem from 1965 to 1993.

71 Rutherford M. Poats (1922–2020), journalist and international development and banking services expert and innovator.

72 Eric Sevareid (1912–1992), American author and CBS news journalist from 1939 to 1977. He was one of a group of elite war correspondents who were hired by CBS newsman Edward R. Murrow and nicknamed 'Murrow's Boys.'

73 Katherine Graham (1917–2001), American news publisher and head of the Washington Post from 1963 to 1991.

74 David Kochav (1927–2010), Israeli economist.

75 Colonel Arie Brown, Dayan's military adjutant who managed his schedule and personal diary; Carl Kaysen (1920–2010), MIT national security expert. Under President Kennedy he helped negotiate a key test ban treaty; Lucian Pye (1921–2008), American political scientist, sinologist and comparative politics expert; Milton Katz (1907–1995) served as ambassador for the Marshal Plan. He acquired such a reputation that he continued to be responsible for international projects. He later served as Chairman of the Finance and Economic Committee of NATO; Mattityahu (Matti) Peled (1923–1995), professional military man who reached the rank of Aluf (Major General) in the IDF and was a member of the General Staff during the Six-Day War of 1967.

76 The important role played by South Korean troops in the Vietnam War is, even today, not properly appreciated.

General Moshe Dayan aboard USS Constellation *(CV-64), August 1966.*

— 2 —

FIRST ENCOUNTERS IN VIETNAM

Saigon
25 July 1966

Arrival at 5.30 pm (local time) via Los Angeles, Honolulu, Tokyo (overnight) and Manila. Press attaché Needham from the American Embassy in Saigon is waiting for me at the airport.

On the way to my hotel (Embassy), he asks to take into consideration the fact that both General Westmoreland[1] and the Ambassador, Henry Cabot Lodge,[2] want to meet with me. I tell him what my purposes are. Most likely, tomorrow I will be busy getting the necessary permits, and my visit will begin the day after that. He has already been told that I want to visit an aircraft carrier: this has been arranged for the day after tomorrow. After that, he promises that I will be attached to one of the units. He will try to arrange meetings with the army commander and ambassador in a way that doesn't disrupt my other plans. That evening, he invites me for dinner at his home. Two other members of the embassy staff are also there—one of them is Paul Hare (son of Raymond Hare, who is involved in 'pacification' activities).[3] The conversation is obviously Vietnam. My mind gets used to the question—What good is the American war here doing? I enjoy the conversation—or rather those who are talking—simple, pleasant and intelligent people. Although they work here—and Needham must be kept busy by briefings—they have no clear answer to the two critical subjects: why are the Americans fighting here, and how can the war be concluded? Even if I pose disagreeable questions all evening, in the end I formulate two issues which gradually crystallize in my mind during the visit.

If America's strategic problem was self-defense, and—in case of war—destruction of the enemy (and its country), it could do so anywhere in the world, in outer space and under the sea. However, events have not worked out this way. If the United States wants to influence—and not merely destroy—it must also do so by partially military methods—in other words presence, constructive non-military activities. propaganda and use (or threat) of force. It must therefore be present, in civilian and military capacities (land, sea and air power), in parts of the world that it wishes to influence. There are two ways to end the war: either the bargaining table, or the war will die down—as things stand now, both are possible, whichever happens first.

Needham lives in a rented house, with servants, a cook, etc. Although this is my first time in Vietnam, I don't feel in the least foreign or surprised. The streets, stores, people, beautiful silk-clad girls—it feels as if I have been there before. Perhaps because all places in the East resemble each other, perhaps the movies, books, newspaper photographs, all perpetuate the image.

Even the soldiers standing at the crossroads behind sandbags are not new—it is a picture straight from the modern State of Israel—Jerusalem, Haifa or Jaffa—even Cyprus, which I visited several months ago.

Those young Americans like Paul Hare involved, both here and in Washington, with 'American imperialism,' emanate confidence and belief in themselves. It seems to me that this confidence and self-belief—more than any feeling of expression of military, economic and political might—is derived from the fact that they know they are not exploiting the countries in which they work.

Founders of past empires probably also felt no guilt, even though they exploited weaker countries for the good of the ruling country. Americans of this generation know that 'exploitation' is bad. However, they really do not exploit; they invest their own money, develop, and give—with no thought for competition to be the candidate with greatest influence in the region adjacent to China.

In addition, these young men recognize, and experience in their daily work America's military, economic and political might. No one can defeat their army. They are confident that they are building, developing and giving of themselves, not exploiting or abusing. This allows educated, liberal, patriotic Americans to complete their work with pride, and see in it a kind of mission, without any feeling of subordination or unpleasantness. Even when the entire world criticizes them, they simply reply: 'our critics don't know what they are talking about.'

On the other hand, there is an inherent contradiction to what they say. In their objection to Communist rule, Americans assert that the Vietnamese really don't want the Viet Cong. They wish to retain their traditional way of life and will not accept changes demanded by the Viet Cong (agrarian reform, central proletarian structure, etc.).

There are two weaknesses in this argument:

Firstly—The Communist approach advocates revolution, obviously founded on a change in their way of life. They cannot agree to Communist rule but still preserve the way they lived before.

Even more important—no matter whether it's the Communists, the Americans, or anyone who wants to improve their standard of living with advanced agricultural technology, industrial improvement, etc. etc.—their traditional way of life will be affected. I know a little about such things from my experience in Israel. It is no coincidence that women in Arab villages have adopted Western dress: this comes together with the tractor, cinema, and newspapers/periodicals. Israeli Bedouins will not be able to continue their traditional Bedouin way of life and, at the same time, reach the living standard of Israelis. Anyone who wishes to preserve the traditional Vietnamese way of life will have to oppose any change. On the other hand, development and progress must come hand in hand with fundamental changes in way of life (not, of course, in the Communist sense)—but in clothes, customs, hierarchical and familial status, etc.

Therefore, when young Americans enlist the Vietnamese people against the Viet Cong—maintaining that communism will damage their way of life, assuming that it should not be changed—they are 'skating on thin ice.' They themselves are bound to be precursors of change, no less than the Viet Cong.

26 July 1966
Saigon

The morning is mainly occupied with paperwork. After being photographed and completing the necessary questionnaires, I leave armed with three permits. The first, as a journalist accredited to American units. Secondly, a similar accreditation specifically from the Vietnamese government. Thirdly, the possibility to telegraph Israel collect, on *Ma'ariv*'s account.

I am outfitted with military clothing—I receive boots, trousers and jacket, hat, water bottle, etc. boots and jacket from the personal store of Major Hunt; however, his trousers do not 'fit' around my belly (Simcha's cooking)! I am not prepared under any circumstances to give it up in exchange for two extra holes in my belt!). So, I must turn to the quartermaster's store. There they 'take me in hand.' They restrain themselves with difficulty when it comes to a submachine gun and grenades.

Before leaving for various offices, the sergeant gives a briefing to me and another American journalist. I don't know what the journalist thinks: I am not carried away by the sergeant's enthusiasm.

He certainly knows his subject: he runs over the map with his finger, correctly pronouncing all the Vietnamese names, even knowing the exact coordinates of the entire region—swamps, mountains, jungle, etc. However, when we pose a few factual questions about the results of the battle, he has no hesitation saying—'Sirs, I am a public relations man and not an information official.' He doesn't hide the fact—that he simply doesn't know.

One element stands out in what he says—technical military capacity—capability of reaching everywhere by helicopter, the speed of speedboats patrolling the river, assistance by aerial firepower, etc. Without properly understanding the issue, he also heaps praise and enthusiasm on the Korean Army—'they have been made into real men.' Their spit and polish, how they jump to attention when an officer passes by! He doesn't lay any emphasis on their military capability—it must be in their blood, not acquired from American leadership. The Korean army is also one of the American army's technical achievements.[4]

It would not have been necessary to mention any of this if it was a 'malady' specific to the sergeant. But there is no doubt that a great many here are drunk with the mighty technical capacity developed by the American army. I think that this plays a significant part in feelings of superiority (which have replaced doubts, misgivings and questions—'why am I here?') on the part of even the most intelligent and well-informed members of the American army.

We have another type of conversation with the publicity colonel appointed by Westmoreland. He complains that Fulbright[5] and his friends are making Ho Chi Minh believe that the American government is weak, because of domestic opposition to the war. He speaks enthusiastically about the achievements of 'Operation Hastings,'[6] which has damaged the Viet Cong by crushing parts of a division before it could begin 'stage 3' of operations, according to Ho Chi Minh (concentration and attack). I ask him what would have happened if they hadn't scattered the Viet Cong in time, allowing them to launch a damaging attack. He has no answer.

'Hastings' did not start as a previously planned operation, but rather a series of escalating encounters by American reconnaissance patrols with the North. The operation was decided on when concentrations of Viet Cong troops had been confirmed. The Americans assumed that the Viet Cong's aim was to gain control of the northern provinces of Vietnam during the monsoon season.

More than eighty Marines, and (at least) 500 Viet Cong were killed or wounded during the very heavy fighting. A battalion of Marines and three

Marine helicopters collided with each other and were damaged (a fourth helicopter was shot at from the air and crashed, killing all on board). The battalion was moved elsewhere, but two platoons were sent to destroy the damaged helicopters. The platoons approached the helicopters, receding away from their battalion; concentrated fire then opened separating them from their battalion with heavy losses.

In total, 1,600 Vietcong were killed or wounded during 'Operation Hastings,' apart from bombings.

Seven battalions of Marines (5,000 men) fought in the battle—apart from reconnaissance units, etc.

In the meanwhile, a comment by General Ky[7] on an article publishing by US News and World Report is published today: 'only a military incursion into North Vietnam can bring about total victory'—in other words General Ky sees no possibility for victory the way the Americans are currently prosecuting the war. It may be assumed that they are not listening to Ky's declaration (although he is the country's head and commands the army!). His reason for pushing the Americans to conquer the north is understandable—but the significance of his negative opinion regarding the war cannot be overlooked.

After arrangements of military permits, I still have an entire hour until the afternoon mail arrives and can obtain a press card enabling me to send reverse charge telegrams to *Ma'ariv*. I use the time to walk around the city, not only to see the city itself, but also to get accustomed to it—its people, weather, Vietnamese voices, etc.

I know from personal experience that a new, unfamiliar world elicits visceral responses that distort correct judgment. Simply put, a place into which one is thrust at night gives rise to imaginary images, although the same place causes no feelings of fear or excitement by day. Humans have well-developed imaginations: a person has only to sit in a red velvet cinema seat to feel that he is in the jungle, only because there are pictures of trees and snakes on the screen. One must get used to the screeching of wild birds by getting used to the area, walking around in it. This is necessary to change the initial image of Saigon into a day-to-day occurrence and decrease the initial dramatic effect of the swarms of people in the streets—on foot, by bicycle or transported by three-wheeler. Men with childrens' faces, women with faces of Madonnas and bodies like Brigitte Bardot, dressed in tight-fitting bright silk pajamas in all the colors of the rainbow.[8] One must get accustomed to the heat, humidity, the sharp sound of the twittering speech. And despite all this, one must be able to judge things, events, people, their words and actions in plain, undramatic, human-secular terms like someone in his own country, a place where he sees things by outward appearance, without being impressed anymore.

During periods of military activity, I have always tried to disregard outside effects—dark of night (comparing it in my imagination with daytime), the noise of gunfire,—to ignore explosions and the shriek of bullets, wounded, etc. If not for this, I would react to stimuli based only upon dramatic outside reactions like light, dark, noise, etc.—thereby missing reality.

I would like to feel my own 'secular' ground under my feet: several hours of walking in the streets would help dissipate my initial feeling of strangeness. But one gets used to it, and then it doesn't divide or distort clear vision.

Towards evening (upon advice of Chechik,[9] I meet with a delegate from the department of defense, called H. Cross.[10] We had met previously for a few minutes and he arranged for us to have dinner together. He promised to try to get Professor Le Van Tho[11]—from the institute of atomic energy—to come as well. Professor E. D. Bergman had recommended that I meet him, and gave me a letter of introduction.[12]

At 8.00 pm we go together to restaurant 'Cane'—Cross, his assistant Alex Jean.[13] Professor Le Van Tho, and me. I tell them the reasons for my visit, and ask whether I could meet with Vietnamese people, to obtain their opinion.

During the evening, the Professor repeatedly tells me that he cannot, in time of war, voice opinions other than the official version. Anyone who does this is in mortal danger. 'You will go home, but I remain here.' Therefore, he believes that I can only meet with 'official people' and hear official opinions. This, in relation to my visiting villages to evaluate the effects of 'pacification.' Anyone accompanying me not voicing the government's opinion—would be in mortal danger. The professor can therefore only introduce me to 'official' Vietnamese citizens, who will also just give me the official version.

The little that he can tell me, is that the Vietnamese people don't want to continue, or even win, the war—only to stop it. They are sick of it, and it is destroying them (the nation) morally more than physically. Men and women know that they can obtain (American) money, not by hard, strenuous labor, but by all sorts of speculative activities (he speaks with great caution and does not specifically bring up the question of women and American soldiers). The nation's character is being destroyed.

The elections would not be free. The entire issue of free elections is very complicated. Many families are divided—part of them in the North—and those there would be punished severely if members of their family voted against the Viet Cong.

The Viet Cong are much stronger than the Americans realize. There are two divisions in Saigon armed and ready (Cross told me this, and the professor confirms it). They are just waiting for a sign of activation. Maybe this will not occur, but as said, they are much stronger and more numerous in South Vietnam than the Americans believe. If real elections took place—the present government would immediately fall: the people have no confidence in it.

General Van Minh would be an acceptable candidate.[14] He is a decent man and recognized leader. (This was the same general whom U Thant[15] had praised: he had said that if he were there instead of General Ky, the government would be acceptable to the Vietnamese people). But he is presently in exile in Bangkok. General Ky knows that he represents competition and will not allow him in Vietnam.

We arrange that, when I return from the north, I will contact him, and we will meet again.

The restaurant is full to bursting—mainly with men. It seems to me that many are French (businessmen). The wine, food and atmosphere—are all French. There is certainly no expression—in the people or the atmosphere—of Vietnamese tradition or the suffering caused by the war. On the contrary: the people, plentiful amounts of wine and expensive dishes are all expressions of a period of war characterized by 'easy money' (for some of the people) and the desire to spend it, enjoy it, and be seen doing it.

No one can—I believe—argue with the fact that Vietnam, today, has been conquered by the United States. It may be that past associations with a conquered country do not apply here, and that a more accurate idiom must be found. But, no matter what, it's clear that what is happening in Vietnam today has not been decided by the free will of the Vietnamese people. President Johnson's literal formula — that the United States is only here to allow its South Vietnamese allies to choose how they want to live—is not tenable. The United States is protecting Vietnam from not being conquered by the Viet Cong but is not allowing the people to live as they wish. It is possible that both cannot be allowed together and that one of the two must be chosen. But even this choice is not, today, in the hands of the Vietnamese people. It may be said that this is only a temporary situation, a war emergency—but its influence will not fade after the war. Vietnam today is attached to a much stronger force—the capabilities, actions and influence of United States power (both civilian and military). Free and fair elections—choosing a government that the people really want—are not, nor will they be held, in Vietnam, now or in the future. The Vietnamese people will not be allowed to decide this basic issue for themselves—whether they prefer this war over the expected peace after its settlement, or vice versa.

It may be that this question and its answer do not coincide with the interests of the Vietnamese people, and that they should therefore not decide themselves. It may be that Vietnam is nothing more than a theater of action for a violent (national) struggle between the United States and

China, with the Viet Cong simply a victim trampled underfoot by forces superior to it in military strength, political influence and economic abundance. Whatever the definitions will be, the formula that 'the Vietnamese people must be able to choose how they wish to live'—even if it is believed through naïveté—is in reality cynical in the extreme.

I have written to the Burmese Ambassador asking whether I can come for a visit there.

Boat patrol
27 July 1966
Saigon

At 7.30 am, I go out to observe a swift-boat patrol on the rivers of the Delta.

We travel by car to the Saigon military airport (the busiest in the world, at any rate busier than Kennedy airport in New York—which is currently on strike …). From there, by helicopter to the naval base. I receive a briefing from the base commander and go down into the boat. The tour is arranged in two or three boats. Their speed is over 25 knots. The boats are made of fiberglass and equipped with different kinds of machine guns and 50 mm mortars.[16]

The boat crew are a pleasant group—an officer and four men. During the night, we navigate by radar and infra-red light. During the day—if there is no fog—we patrol visually. On the way, we stop twice: once, for a routine search of a convoy of boats transporting firewood, and once for a weapons exercise.

The search reminds me of the British military police in Israel during the Mandate, at a time of good relations. The officer stops the Vietnamese motorboat which tows another eight boats behind it. His boats are loaded with tree trunks and branches of wood from good trees.[17] Three Americans enter the Vietnamese boats, examine their papers, and open the container covers. Of course, they find nothing suspicious. I truly don't think that there is anything else in the Vietnamese boats. But if there was—they would have hidden it under the wood, and who (which American) has the patience—and justification—to unpack a load of wood to see what is underneath? Simply put, if all the patrols were like this—the Viet Cong would have no difficulty secretly bringing in arms and ammunition.

By the time the Americans have opened the container covers, I have removed the covers from the pots standing on top of the coal-burners, boiling.—Our meal is cooking—rice, fish (sardines), and some kind of vegetable dish. It smells good.

The river echoes with the noise of transport boats, fishermen, steam, paddles and engines. Transportation by river is certainly a great deal cheaper than by land. The Delta area is cut through by rivers horizontally and vertically, and by tributaries which connect and intersect with each other.

The boats do not only carry passengers, but also the sailors' families, who bring up their children and live there—from birth to death. Each boat has a kind of 'living room,' 'kitchen,' etc. One can mock them—but the heating element is lit, and a meal is cooking on the coals.

There are, naturally, plenty of children, who probably don't go to school; but they know the river well. American equipment is excellent, suitable for young American sailors. It is a pleasure to observe them and talk to them (the officers). They are very interested in hearing my opinion of the war, in other words whether they (the Americans) are doing the right thing—from a world-perspective and not specifically for American interests—by being here. In their opinion—yes. The reason—confrontation with China—on the sea and on the ground. Every time they look at the land, they say: 'what a country!' Clearly, they are happy to be in the navy.

The land on both sides of the river is covered with bushes and low trees and tall grass, through which cut or broken tree trunks poke out. The area closest to the river is swampy, puddles sparkle between the bushes. Marching in such terrain doesn't attract me one bit.

During the second half of our patrol, it pours with rain. The men take their shirts off—'a free shower.'

From time to time, we see the South Vietnamese Army camps on the riverbanks—one to three platoons—a sign that they are not concerned about a real attack.

Since the American Civil War—this is the first time that the American Navy performs river patrols.

The impression that this demonstration of immense power (the many units on patrol) has on the population must be considered—It appears to them that the Viet Cong are so strong, that all this is needed to defeat it. (Also, the Vietcong are deceiving them and crossing their boats). It cannot be discounted that all this is useful to strengthen the impression of Viet Cong power.

I would not suggest leaving the river, but rather the establishment of routine military police patrols with a much smaller force (perhaps 10%) and, in certain places, establishing sea and air reinforcements, which can be summoned by radio.

To sum up, my first 'day of fighting' with patrols: this is not the way to reach the objectives. It isn't the way to fight guerillas, nor does it achieve the desired effect on the population (convincing them that they can obtain security from South Vietnamese forces). Fewer patrols, by the Vietnamese Navy, would be better, with Americans ready to come to their aid when needed.

In actuality, I am probably suggesting that Americans return to the role of 'advisors' for river patrols.

In any event, I don't believe that it is possible to prevent infiltration of men and arms, given the conditions of the water and region. If searches were done correctly, they would close the river traffic. Can each cargo be turned upside down daily? Absolutely not. Therefore, the question is not of negligence or not using waves of mines, but the impossibility of boarding and properly examining each boat.

In such a situation the determining factor is not preventing infiltration, but the Viet Cong's inability to defeat the Americans, even when their rifles do reach the South.

The Viet Cong obviously could—and still can—attack each small village, even temporarily conquer small towns in different places. But, if this would change to fighting with the Americans, and shedding blood of the locals—it certainly wouldn't do them any good.

After the patrol, we move to a larger coast guard (sea) ship. I attend a regular firing exercise, and an alarm drill.

The routine (sea) patrols tire the commander officer and men out—day in and day out, without mishap. We return by helicopter.

American arrangements delay us at the airport for 2 h; for some reason, the car doesn't arrive. This is bad luck for me,—I am extremely tired, after an almost sleepless night and a day on my feet (in the boat), etc. I cannot say that I accept this small mishap philosophically.

At the end of the patrol, the officer accompanying me tries, for the sake of naval 'public relations,' to find out my conclusions, and keeps on asking me what my impressions are. I keep on delaying, telling him I didn't know. He decides to 'assist' me, and formally announces that he too has long pondered the question, whether the United States should be here or not. After seeing the situation upfront, he has arrived at a positive conclusion: American presence is necessary. Likewise, the young men are doing 'excellent work.' Anyone who doesn't see the expression on my face after this announcement—cannot properly evaluate the patriotic devotion of these public relations officers ...

Aircraft Carrier
28 July 1966

On the deck of the aircraft carrier *Constellation* from the 7th Fleet. This is the biggest ship (of a total of six). We arrive at noon—straight to the mess. I can't see the landing because I sit tied in and effectively immobilized. The shock of the sudden stop is not very strong.

At 3.00 pm—take off of a squadron for an operation. A short time after that—landing of returning squadron. The circuit usually takes a total of 1 ½ h.

The operation is breathtakingly impressive. Take-off and landing on the short 'runway,' organization of reception of planes on different decks, effect of seeing 5,000 men working on the ship in exemplary fashion—all this, despite what I already heard and read, exceeds all expectation.

We are not informed exactly where the ship is. But, from all that we can see, we are about 50 kilometers from Hai Phong Bay, in other words from Chinese and North Vietnamese territory. In answer to my question, they tell me that enemy planes have not, up till now, tried to approach them. In any event, the ship has sufficient firepower to defend itself (there is another aircraft carrier nearby—with mainly helicopters—for a total of four ships, if he is not mistaken).

They show us everything but refrain from answering operational questions—especially the objectives they are currently bombing.

During a conversation with one of the pilots (a squadron commander), he complains bitterly about the heavy anti-aircraft fire they encounter, to which they have no response. At high altitude, anti-aircraft rockets work efficiently; at low altitude, conventional light weapons fire comes into play. I get the impression from the pilots' words that they have no way to disrupt the trajectory of the (SAM) missiles.

They emphasize the difference between the war in South Vietnam and in the North—according to them, these are two separate theaters of war. In the North they are fighting with equal participation of air-force and navy. They speak mockingly about General Westmoreland's slogan 'search and destroy.' In the South—they search; in the North—they destroy.

There is no more striking example of the 'Americanism' of this war than aircraft carrier operations. The South Vietnamese government has nothing to do with this. The names are interesting: they called the Korean War a 'police action'—the current one is a guerilla war—with aircraft carriers ...

A seasoned senior officer from the 7th Fleet emphasizes the following points in briefing discussions:

1. The American fleet is sovereign of half the world; there is no more lethal force anywhere.
2. About 90% of the world's town- and city-dwellers live near the sea; the navy is capable of dictating terms in these areas, which comprise most of the civilized world.
3. The navy can operate immediately: it is always ready for action.
4. Its strength is stable and capability continuous. It carries its own provisions, and has its own method of logistic flow, independent of foreign elements (they neither give nor receive anything from Vietnam).
5. They have no security concerns—spying, etc.—its men are locked in the ship, under supervision.
6. They have their own land forces—the Marines.
7. The Marines continuously build up their power, through continuous training, and therefore have the force of 'continuous power projection.'
8. Great work output. The men have no alternative, to pass the time.

During my visit, I ponder: there is nothing more dramatic than war, and no aspects are more dramatic than those that constitute it—from death and destruction to idealistic self-sacrifice. However, no other issue must be freer of dramatic or secular influence (this applies from an ordinary soldier to the commander in chief).

Dinner with Rear Admiral James R. Reedy,[18] Commander, Attack Carrier Strike Force, U.S. Fifth Fleet[19] (*Constellation* is his flagship). They have informed him (from Washington?) of my arrival. He is surprised that Westmoreland did not invite me for a visit. In the beginning, conversation is strained, but, after a while, we connect. We find a common acquaintance—General Fergusson, who was Wingate's deputy in Israel in 1937, and is now High Commissioner of New Zealand.[20]

The atmosphere between us becomes freer, and I ask him 'how the war in Vietnam will end.' He criticizes the land forces: the American army should establish a permanent front on the seventeenth parallel and cut the South off from the North: the war will never end, if they keep jumping from place to place. I doubt that his recommendation is good; but apparently, he is counted amongst the many who don't see how the Americans can force Ho Chi Minh to peace talks—the way the war is being prosecuted now.

The military policy, that the problem lies in America's place to project strength in the world, is also prevalent amongst the officers. No one can overcome it; the Chinese are a fiction (from the point of view of force). The problem is not quantitative, nor manpower related. If the United States must fight on for 10 years, it is ready to do so. No one will mention South Vietnam, General Ky, etc. later—they are of no interest to anyone. The question is: America against the entire world (a position of hegemony).

To the question why the Soviet Union do not have aircraft carriers here, the admiral answers that the Russians certainly do not have the ability to build ships like these. Up till now its strategy has been built on defense, whereas America's strategy is attack (in clearer terms, military activities overseas).

Montgomery's name comes up during conversation. Admiral Reedy served in NATO while Montgomery was its chief, and his wife was Montgomery's secretary. As the admiral describes it, Monty was a lonely man (after his wife's death)[21] and talked a lot with Reedy's wife (while Reedy was still a captain). All day (even at mealtimes) a photographer from 'United Press' stays glued to my side. He has received instructions to 'cover' me—with photographs. Apparently, the call has gone out to all United Press cameramen, because a young man comes up to me at Saigon airport and says that he will accompany me for 3 days, for photographs (afterwards they don't take him to the carrier by plane). The types and appearances of journalists milling around here—are fantastic! A sunburned red-head from United Press Saigon, beard in an advanced stage of growth (long stubble), eyes red from a worrying lack of sleep, dressed in battle fatigue—a real piece of the jungle—apart from the red head. The jungle would not raise such a flower even during a long drought ...

29 July 1966. Aircraft Carrier *Constellation*

I get up at 3.00 am to observe night-time operations. The ship's activity never stops, day and night; every 1 ½ hours a squadron of planes departs and returns. During the night, the ship is flooded with light. During the day, it activates a strong wind (counter) to receive planes. Two planes are off target, turn around—and land. During take-off and landing, one rescue helicopter (they are called 'angels') circles over the ocean in case an accident occurs, and it must hurry to a wounded pilot's aid.

The ship carries about 100 planes and about 300 pilots. It is actually a factory (heavy industry) for aerial warfare, which manufactures a series of war operations every hour and a half. The ship can protect itself from the air, sea, land, space and under the sea (submarines, torpedoes, etc.).

There was a Mexican menu at last night's dinner (who knows why?). I can't say that it got me excited. Of course, everyone to his own taste, but tonight it was not a subject of argument—just eating.

Breakfast makes up for last night's dinner. The ship's daily 'newspaper' lies on the table, summarizing current operational developments. Almost nothing in the South (the impression—the Viet Cong are not active, and the Americans don't know where to attack them). Bombings continue in North at full speed. Amongst the objectives mentioned: railroad tracks and freight cars. One cannot doubt their ability, when I say that the Americans have decided to pressure the North Vietnamese to the end, and that they will not stop until all road and rail traffic, not just military transports, cease. They want to force Ho Chi Minh to submit to an imposed peace settlement. My impression is that they are not fighting against infiltration from the North, not even a guerilla war or a war against Ho Chi Minh, but rather an American war against the entire world. Their goal is to demonstrate their power and strength of purpose to everyone (mainly Britain, France and the Soviet Union) so that they know that, when America enters a war, no one can stand before them.

By the way—in Israel, when one speaks about 'American guarantees,' one must know the meaning of 'American military aid.' This is not placing the 7th fleet under command of the Israel Defense Forces, but just the opposite—transfer of Israel's sovereignty into United States hands. To protect the independence of an ally—they first must hold it in trust.

All this does not relate to the character of the people themselves: they are liberal, polite, pleasant and easy-going. As individuals—they are 'golden.' But all up to a point: until anyone shows 'disrespect' for United States power. There is no elasticity in this issue, even in conversation. They do not hide their feeling that they are stronger than the rest of the world combined. They don't care what others think of them and don't need their help.

There is some sort of problem with the plane which is supposed to take us to Da Nang, and we will only leave here at 3.30 pm (perhaps the reason is—the location of the ship, which is probably somewhere near the Chinese coast).

Endnotes

1 See Chapter 1, note 50.

2 Henry Cabot Lodge, Jr. (1902–1985), American diplomat and politician who served as United States bassador to South Vietnam 1963–1964 and 1965–1967.

3 See Chapter 1, note 59.

4 Patronizing. The South Korean contribution to the American war effort in Vietnam was significant. They were already good soldiers when they arrived.

5 See Chapter 1, note 41.

6 Operation Hastings (late June and Early July 1966) was an American military operation in the Vietnam War. The operation was a qualified success in that it pushed the People's Republic of Vietnam forces back across the Demilitarized Zone.

7 See Chapter 1, note 47

8 Dayan had a well-known eye for a pretty woman.

9 "Chechik" refers to an alias used by a CIA officer. The real name of the person associated with this alias is William (Bill) Chechik. He was involved in U.S. intelligence and defense affairs, and his interactions with Israeli officials were part of broader intelligence and strategic discussions between the two countries.

10 Harold Cross was the Director of the Office of Near Eastern and South Asian Affairs in the U.S. Department of Defense.

11 Lê Văn Thọ, Vietnamese atomic physicist working for the South.

12 Ernst David Bergmann (1903–1975), Israeli nuclear scientist and chemist often considered the father of the Israeli nuclear program.

13 Alexander "Alex" Jean, Assistant Chief of the Near East and South Asia Division in the Office of International Security Affairs at the U.S. Department of Defense. His role involved working closely with key officials on matters related to defense and international security in the Middle East and South Asia.

14 Dương Văn Minh (1916–2001), South Vietnamese politician and a senior general in the Army of South Vietnam and a politician during the presidency of Ngô Đình Diệm. In 1963, he became chief of a military junta after leading a coup in which Diệm was assassinated. Ironically, he was the last President of South Vietnam in April 1975, two days before surrendering to North Vietnamese forces.

15 See Chapter 1, note 44.

16 Dayan seems to have made an error here: 50 mm mortars were used by the Russians. The US Navy used 81 mm mortars and .50 caliber machine guns and 40 mm machine guns to arm its small patrol craft in South Vietnam. Some also had a 60 mm mortar.

17 The quality of wood from Southeast Asia is very high.

18 Rear Admiral James R. Reedy (1910–1999).

19 This is an error: The 7th Fleet was active in Vietnam (as Dayan notes later).

20 Brigadier Bernard Fergusson, Baron Ballantrae (1911–1980), British army officer and military historian. Dayan spells his name wrongly as Fergelson.

21 Montgomery's wife died in his arms from septicemia following an insect bite and amputation of the infected leg in 1937.

General Moshe Dayan aboard USS Constellation *(CV-64), August 1966.*

3

WITH UNITS IN THE FIELD

The Marines
2 August 1966
Da Nang

During the past few days, I have been in the field with units during an operation. The corps commander is Lieutenant General Louis Walt,[1] the division commander Major General Kyle.[2] One of the battalion commanders is Brigadier General English[3]: the present one's name is, I believe, Fields.[4] While I am with the units (company staff), I join reconnaissance operations with subunits—platoons. During a great deal of walking, I make only one copy of my notes. I send my records with John Snyder[5]—the red-headed United States cameraman who accompanies me: I hope they arrived safely in Israel (to Moish Perlman[6]). I will not try to rewrite this portion. Tomorrow morning at 4.30 am I must depart for the airborne cavalry and will hardly have time to write my notes for today and yesterday.

A wounded animal rampages in the jungle and drives the unit crazy; they rain fire on it all night.

Yesterday morning, the battalion receive orders to withdraw, and I join a platoon led by Lieutenant Charles Krulak—2nd Battalion, 7th Marines (2/7), 1st Marine Division. I travel on foot, downhill of the battalion—only a few kilometers distance, but a steep, slippery, difficult descent. At times we run in standing water, at other times in a strong current.[7]

From the battalion, we leave in helicopters, which 'lift' the entire battalion. Helicopter artillery consists of 105 mm howitzers, each cannon suspended under the machine. We are transported to where the division is concentrated. From there I am sent—again by helicopter—to headquarters, home of the corps commander, General Kyle.

Kyle's headquarters in Da Nang are at the top of a mountain—a fantastic view. General Kyle is waiting for me at the door of his house—like the other generals, he treats me with exceptional friendliness and hospitality. This time I badly need a real dwelling, after three days (and two nights) in the field in rain, sweat, bombings which come and go ceaselessly.—A hot shower, good meal, bed with a pillow, are most welcome.

My host for the evening is General English (Kyle is on a visit to Walt). I eat and fall down on the bed sound asleep. The next morning, Kyle invites me to his command's daily briefing—a real briefing, not just for the benefit of the guests. Staff officers present their reports. In the interests of brevity and practicality, there are almost no questions.

The interesting parts are—the reconnaissance encounters. 'Operation Hastings' has been completed or is winding down.[8] However, reconnaissance units are still active in the area, with many daily encounters—mostly resulting in several (two to five) wounded Viet Cong. The Viet Cong are not attacking anymore—perhaps they have no way to defend themselves. Each time they are exposed, they are hit with accurate artillery fire. Reports describe many groups of Viet Cong, discovered while walking, on the other side of the river; or else they themselves open fire on the Americans—in which case artillery is immediately called in, and destroys them.

The Americans also have casualties—by Viet Cong fire, or mines.

After the briefing, General Kyle asks me what I wanted. I say to be attached to a reconnaissance patrol. For the first time, I am refused. Kyle chuckles, and says he knew I would request this, but he is concerned about agreeing. The unit gets split into bands of four to five men, who remain in the area for two three 3 days. Many land on mines, with heavy losses.

We agree that I should return to General Walt's headquarters, to reach the 'cavalry' as soon as possible. I meet Platt, General Walt's chief of staff.[9] My plan is to join General Walt in his tour of two sections. The first is a hospital ship, the second a group of villages where 'civilian pacification' is being carried out.

The hospital visit is impressive—America! America!

To appreciate the enormous power of the United States, one must see the resources, equipment and conditions provided to its soldiers: from battle rations (most have chewing gum, ways to make hot water, cocoa, etc.) to hospital, with the most modern and advanced equipment. Of course, I understand nothing of this, but I believe the ships' doctors when they tell me that their equipment is amongst the most up to date in the world.

The very low mortality (about 0.5%) is mainly due, according to the doctors, to the helicopters (that land on the ship, as ours did). The use of helicopters has brought about a tactical revolution in the field, and a fantastic advance in services rendered. The wounded now arrive at first aid immediately and are very rapidly transported onto the hospital ship. Men often owe their lives to these helicopters. About 1,700 helicopters are available. An hour's travel time, an hour for treatment. The number of wounded who cannot return to the front because of psychological trauma is also very small (2%): the corresponding number during the Second World War was 25%, and 15% in Korea. Perhaps this is also because the Marines are all volunteers.[10]

We lunch on the ship and then go out to the villages. We visit two villages. Marines are responsible for village welfare. As explained to me, this region has already been 'pacified' (The morning briefing—from police sources—relates that some villages are completely under Viet Cong control, others controlled by the Viet Cong only by night). The region (it can especially be clearly seen from the air) is cultivated (mainly rice, some fruit). Everything is done by hand, in tiny plots. From this point of view, there is a huge difference from mountainous areas. There are clearly three kinds of wars here. In North Vietnam, only by air; in the unpopulated mountain areas, where there is unlimited use of all types of weapons; in the plains—thickly populated with villages, one next to the other.

There is not a single male in the village (even boys of 15–16) who is not either in the Viet Cong 'in the mountains' or in the army of the South, or else has died (or been killed). The village elder—a forlorn old man—has agreed to return to the village and walk around in it, but he spends the night in the Marine encampment. During the past few weeks, there have been no clashes, and not a single South Vietnamese soldier has been in this area—not one. The American soldiers are filled with enthusiasm for their work. I believe them when they say they are doing their job—the job of war—with all their hearts. In the mud and rain, with dripping sweat, and water that covers the plains in endless creeks and tributaries. They bring in tanks and 'Ontos' (six barrels of 106 mm with two recoilless rifles, hugely powerful: striking distance up to 2,000 yards). Mortar placements are surrounded by sacking. Communication with artillery, helicopters and planes. The soldiers themselves are healthy and work tirelessly.

Their relations with the villagers are the epitome of decency. The commander points to the groups of beautiful young girls—standing on the side of the road ogling the soldiers—and says that the Marines are not to touch them. I say that they must be angels: no, he replies, they know the punishment for this. Additionally, I am not surprised to hear that, in their spare time, the Marines have made school tables and furniture from Army wood. They have created order in the village and set the population to cleaning the (main) streets.

Everything looks good on the surface. However, when the commander indicates that 'pacification' has been achieved—he may well be right, but this is far from 'normalization.' I have no doubt at all that the Marines could overcome most guerilla-related events.—with the aid of binoculars, air reconnaissance and combing the roads for mines with tanks. (This morning, however, a hole is made in a tank; the Viet Cong have gotten wise and placed an American 155 mm shell under the mine—the resultant explosion penetrates the tank's armor). I can well imagine a company or platoon of Marines, with their immense firepower, technological assets and soldierly devotion—would succeed (more or less) in bringing quiet to the area. But this will require American power.

As far as local forces are concerned—there is not a single male in the villages that we visited. Even in places where they were present—they are a long way from being able to fight the Viet Cong or control the area.

General Walt understands all this: he speaks of a period of 10 years, during which the Marines must protect Vietnam, be present in the plains, and continually bomb the roads leading to the mountains.

I tell him that, in my opinion, the Viet Cong must remain in two places. Firstly, in the mountains, to guard their communications with North Vietnam (the 'Ho Chi Minh Trail'), and find a place to hide, concealing their retreating units in a safe place (this could be done in the mountains). Secondly, in the plains, to obtain provisions and information and ensure no loss of political control and influence over the population.

I am not sure but that the Viet Cong have agreed to the fields being cultivated and planted with rice, to allow them to obtain the harvest—when the time comes.

Amid the idyll—in the village center—are several homes burned down and destroyed by Marine mortars or cannons directed onto the source of attacking fire, even if this was a South Vietnamese village in the middle of the area.

When we return from the village to corps command, we see below us an entirely different vista; a calm sea with sandy 'Riviera'-like beach. Strips of plain, 20–60 kilometers wide, parallel to the coast, furrowed with rivers and flooded areas winding into them, creating an endless number of prominent islands. In-between—agriculture: a garden of green rice, cut up into small holdings; square upon square, surrounded by embankments and trees, and villages of a couple of dozen houses. In the fields—Vietnamese women dressed in large hats and black pajamas, bent over rice plants in water-flooded plots. On the other hand, on the small hills protruding above the plain—an army of Marines: mortars, cannons, tanks, jeeps and caterpillar tracks—an army at war.

Air activity is no less intense than on land—warplanes, bombing and resupply, and endless swarms of helicopters scampering from place to place. There is no doubt that helicopters have not only facilitated the war, but also determined new tactics. The American commander can now (in Vietnam, where there is no air opposition) mark any place on a map with his finger and say: 'that is where I want to be,' and land where he wishes, with equipment and supplies, without a road, river bridge, or jungle trail, with giant grasshopper leaps. If there is any high, steep mountain in the area—the helicopter immediately lands an observation unit, armed with powerful binoculars, on it. And, of course, limitless artillery and airpower are added to the force-ratio. Any Viet Cong rifle fire is met by a rain of shells and, if necessary, aerial bombardment. The planes also 'tear' roads into the jungle with a consecutive sequence of bombs, burning the elephant grass with napalm, etc. The fullness of American power must be seen to be believed—from the aircraft carrier to the hospital ship. From battle rations to operational supplies, sophisticated cameras, radar, etc. And everywhere—helicopters, artillery, and communications equipment which direct unerring fire. Such strength will surely destroy the Viet Cong in any encounter. And yet, up to now, most contact initiatives have come from the Viet Cong. The Viet Cong have only two possibilities: to be overwhelmed, or change to tactics of guerilla warfare in very small groups, with numbers of men reduced to a minimum, etc. However: this will not only lead to Americanization of the war, but also prevent Vietnamization of the peace. A true Vietnamese government cannot rise in the shadow of American power.

War has become the main issue in Vietnamese life; whoever decides it, decides the country's life and policies—all the rest are separate. It is enough to see the children trained in their wretched village schools, the 'enthusiasm' with which they describe their village, as the 'teacher' has instructed them, to visiting officers, and the shining faces of the

Marines—to grasp the enormous distance between military victory—which the Americans can achieve—and normalization of life here.

If only I could speak to the people of Vietnam … but, as the professor introduced to me by A. D. Bergmann[11] said—everything gets bogged down on the one single answer: 'we are not able to speak freely. I cannot ask one of my friends to go out to the villages with you. Go yourself, but if he does, he will be severely punished.'

It is only possible to assume what the people really think and mean, if they could speak freely. The young men of South Vietnam in Viet Cong divisions in the mountains are certainly not fed pro-American information … .

The question is whether it is in the power of the American Army to motivate the young men of South Vietnam to fight against the North, or at least defend themselves against them. Whether the women of the families whose villages are 'pacified' by the Americans—by removing the Viet Cong from the area—would call their husbands and sons to return, and whether they would listen to them, abandoning the Viet Cong.

All this without addressing the question of who or what is right, but rather who and what will get what he wants.

It may be that, if the population will have to choose between the Viet Cong and the American Marines—they will choose the Marines; because of the security, abundance, etc. that they bring to the villages. But the Marines will not remain in the villages forever. The United States will not turn Vietnam into a colony (or will they?). If the choice is between the Viet Cong and the South Vietnam authorities—the latter have not yet shown their strength. It may come to the simple choice between peace and war. But I am not sure whether the population do not see the Americans as the proximate cause of the war.

The (Airborne) Cavalry
3 August 1966
Pleiku

With the 1st Brigade of the First Airborne Cavalry

Leave Da Nang at 4.30 am for the airborne division.

My last dinner with the Marines. Speeches. They give me their insignia as a memento—and I pass on to another unit.

I reach the 'cavalry' at 8.00 am, after a night almost without sleep. The assistant commander is waiting for me in the field—Major General Wright.[12] He briefs me in his office, followed by a more detailed briefing from his staff officers, on division structure, equipment, and their special method of deployment. He also gives me written material (he says: 'McNamara[13] instructed me to show you this and not hide anything from you.').

The division has 450 helicopters. The heaviest of these can lift a 155 mm cannon, or a 105 mm cannon with ammunition, or a jeep with men, a fully equipped operating theater, or headquarters trailer, etc. A jeep is only used for complicated communications equipment. Everything else is transported on foot or in other, lighter helicopters.

The unique abilities of this over other units—such as the Marines—transported by helicopter are:

1. Expertise—landing in places where others cannot (they have all sorts of specialized accessories—such as a bulldozer transported by air—which allows them to land forward strike groups and prepare any place for a helicopter landing. They have 'integrated' communications with the helicopters—1,500 men work in maintenance; 6 hours work for each hour of flight!)—this allows them to 'give a ride' to entire, fully equipped battalions. In a short time—4 h after notification—the battalion is already in the new location.

2. They are capable of special tactics—grasshopper leaps—which they use to land battalions, and advance them from place to place, as needed.

There are nine battalions and divisional units in a division. At the moment they are in the middle of a large and special operation—'Emerson.[14]'—One of their brigades encounter large Viet Cong forces in the vicinity of Pleiku, so they have decided to begin the operation by adding reinforcements. This is how most of large operations have begun, including 'Hastings' and 'Crazy Horse.'[15]

I ask to be attached to the fighting divisions, but during the morning and noon hours fog is so bad that it is impossible to fly (although they say that helicopters fly 90% of the time throughout the year).

We fly above the areas of 'Crazy Horse,' and visit one of the battalions at the top of the mountain.

The areas in which helicopters land and drop battalions of men are quite unbelievable. There are no limits: right at the peak of a mountain, on a hanging rock, etc. Not only that—but progress is made by giant helicopter leaps.

Obviously, all this is possible because the Viet Cong have no air power. If they had—it would be another matter. But there is no 'if' in war. If there was no jungle here, etc.—war would be conducted differently.

Since the time that the division has been in this vicinity, the population has returned to the area and started to work the fields (about 50% of the nearby fields are cultivated). The soldiers are occupied with civilian development here as well: agriculture, school renewal, health, etc. etc. I am told that it would take 10–20 years before a local administration capable of taking work responsibility would be ready. The camp is well and truly fenced in against the Viet Cong by mines, watch towers, anti-personnel radar, illumination, etc. They keep 1,000 men there, which is 20% of total troop strength. The fence is 23 kilometers long. Helicopters in the middle of the camp—right in the eye of the division, with camps and units all around.

The view is—fantastic. High, sheer mountains, cut by deep, narrow valleys. All had been cultivated—but now, most are abandoned. Rivulets stream in from all the ravines, becoming rivers in the plains. The main road (Route 19) is active and full of traffic. The Viet Cong have blown the bridges, but each time they are (provisionally) repaired. The trees covering the mountains are deep green—one huge mass of growth.

The French suffered heavy losses here. They show me a cemetery with the graves of 5,000 men. (The famous battle in which 1,000 men from a French armored column all perished at the same time.)[16]

The fog lifts during the afternoon, and we travel by plane (with the general and his assistants, who have arrived from Washington) to Pleiku—where division headquarters are located. Here (in the pouring rain), I receive a briefing on what is going on. There has been almost no enemy contact during the last 24 h; they believe that Viet Cong units are advancing into Cambodia, where they can find refuge, assistance, food, etc. The Americans are not allowed to cross the border and pursue them. Up till now, there have been forty-four Americans, and thirty-two Viet Cong, fatalities but Viet Cong losses from bombings in the jungle have probably been higher.

In the meanwhile, it has become dark. They ask what I want and I revert to my request: to be attached to a fighting unit. They suggest that I join the staff of the third brigade, which is in reserve. I tell the staff officer (Colonel Lee) that it is not worth my while to have come all the way from Tel-Aviv, just for a staff or brigade visit. I am not requesting reserves or a brigade—I am requesting a company on active duty. I will not complain if they do not agree—after all, I am a guest, and they are waging war. But if this is the case, I will request to leave tomorrow.

I hope that, nevertheless, the matter will be arranged. Tomorrow will tell...

War is war, and staff headquarters are staff headquarters. Dinner (for which they apologize) starts with liquid refreshment, scotch whiskey, martinis, and many other alcoholic libations—then the meal itself, with appropriate rations, followed by coffee and brandy.

Our relations are very good. Some of the captains have read my book about Sinai:[17] they use it for training purposes. They give the general from Washington and I their best accommodation.

It may be possible to visit 'civilian operations' here as well. In any event, if they attach me to a fighting unit tomorrow, I will stay here until the end of the battle, even after the time scheduled (in that case, I will inform Saigon that I will be late, and unable to attend the dinner to which I have been invited).

This region—comprising the passage from the 'Ho Chi Minh Trail' to the hinterland—is clearly one of the most critical in Vietnam: Here, even at high altitude, there are also flatter plain-like areas. The Vietnamese population, who usually live in the coastal regions, extend inland here, up to near the Cambodian border. From a military point of view, this unit of airborne cavalry is something unique. If only they will allow it—it will be important to see it in action.

From the little that I have heard here, the division's activity resembles that in other places. The initiative for operations usually comes from the Viet Cong who could refrain from attacking, if they wished. From the time when the Americans make contact and bring in artillery and helicopters—the Viet Cong are immediately struck very hard and start to leave the field. If I have seen correctly: in the absence of a political solution—a condition of 'non-peaceful coexistence' will develop. The Americans will carry out large-scale operations, and will, with each contact, cause disaster to the Viet Cong. The Viet Cong will go over to stalking, sabotage and short-lived ambushes (during the monsoons). They will split up into small groups, hide in the jungle, amid the population, across the border, etc.

4 August 1966
Staff headquarters of the First Airborne Cavalry Division

Commander—Major General Norton.[18]

Chief of staff: Colonel Wolfe.[19]

Perhaps I will remain here longer than intended. The division is preparing a large-scale operation. General Westmoreland[20] has been with them for an entire day. The division commander suggests that—if I want to see the division in action—we stay until after the Sabbath (today is Wednesday). Many journalists have also arrived here—surely not by accident.

My understanding of what is being done here increases daily, with explanatory help from the brigade commander—'my name is Barnes [*sic*]'[21]—I tour the area with him by helicopter—right up to the Cambodian border. We land at a 'special forces' camp, and at another brigade headquarters.

More explanations about different kinds of aid (strategic air command lands the men on an area 4 square kilometers, 'peeling off' bits of mountain and jungle). This gives an idea of American strength: firepower, mobility, ability to land anywhere. The only area of American strength that does not impress me is—intelligence. I am not sure whether they are aware of this—maybe they can reveal the location and number of the enemy by technical means (such as inspection from this mountain).

During dinner (at his invitation), I have a lengthy conversation with the general. For some reason, he decides to devote time to me. Is he really interested (they must have told him about me), or did someone 'on high' tell him something?

He starts our conversation with a formal declaration of his two aims: 1) to help the Vietnamese people; 2) to advance the airborne infantry (give them new fields of experimentation and training). He then asks me for my impressions: I tell him that it is too early for me to say, but—

At this stage, the Americans are fighting their own war against the Viet Cong: not help to Vietnam, nothing else—quite simply, the United States against the Viet Cong. How they reached this point—out of a desire to help the Vietnamese, protect the Geneva agreement,[22] or any other reason is unimportant. They will not stop the war, even if the good of Vietnam demands it (and who will decide?). They will not turn this issue over to the UN. At this stage, it is solely a matter between the United States and the Viet Cong—and them alone.

The war's objective is therefore quite simple—to defeat the Viet Cong. Maybe this isn't so simple (for example—they may splinter rather than defeat it, causing fragmentation of the war—ambushes, mines, encounters in small units, terror attacks, etc.). But the goal of the war (because it is an American war) is clear and simple—defeat the enemy.

For this purpose, they are conscripting and concentrating a mighty force; they fire on every enemy sniper with artillery barrages—and rightly so. Firstly, it is in their (financial and industrial) power to do so. Secondly, they spare the lives of their younger generation: young lives are the dearest thing, because they are irreplaceable. I therefore do not reject Americanization of the war—it is inescapable. The Americans cannot play 'second fiddle' to the South Vietnamese—or anyone else, for that matter.

But they are doing something else. I have not yet studied it and it wasn't obvious to me at first glance—Americanization of the peace. Doctors, teachers, administrators, (positively-directed) will teach the children baseball, make them scouts—all this without pay. Vietnam can—like any other nation—receive outside help, but not patronage. Its progress must be organic and independent—with advice and assistance, not dictation and humiliation, which will not succeed, and is unjustified. It seems to me that Vietnam is being 'adopted,'—but I have not studied this question closely enough.

My criticisms surprise the colonel: he wants to talk to me more and asks me what the solution (to lack of local administration) is. I reply that I don't know, but Israel has a good name, as a nation that assists developing nations. I myself have been deeply involved with this. I know how complicated and sensitive this all is, and that 'all that glitters is not gold.'

I remain with Barnes [*sic*] and his commanders until noon, and lunch with the battalion. I want to visit the nearby village, looking for a translator, and am told that the battalion's doctor—named Aharonson—speaks Vietnamese. He is a young Jew who tells me that he had tried to learn Hebrew for years—and failed. But he learned Vietnamese in 3 weeks. (Afterwards, Barnes [*sic*] tells me that he is about to marry a young Vietnamese woman; that is the reason for his fluent Vietnamese).

In any event, at noon Aharonson tells me about his work with Vietnamese doctors. It is exactly as I thought—they let the American doctors do their work, but they themselves concentrate on private practice (after office hours). Of course, American doctors can direct hospitals and provide medical care here. But what they are doing does not advance Vietnamese medicine. There is no chance that, in a year or two, they will easily be able to pass the work to the hands of local doctors.

The afternoon clears up a bit. We fly—above the French tea plantations—to the Cambodian border. There are about 700 mountain people, all working for wages, in the 'special forces' camp; with them, are twelve Americans. Of course, an airfield, artillery (105 mm), etc. as well. Everything looks too static. Such a border camp (and there are many like it throughout the country) is doubly important. It serves as an operations base (also a little for patrols and intelligence), and to bind the locals to the American Army, instead of them joining the Viet Cong. I have a negative reaction to the heavy fortifications—no end of barbed wire, sharpened bamboo sticks, belt after belt of mines, deep bunkers, etc.—a psychology of defense.

In the evening, I attend a press briefing by the general; after that he takes me to the war room, organized, and humming like a beehive. Dozens of telephones and teleprinters; the general walks around in their midst—tense with war excitement. It is a little strange to see dozens of American officers chomping on cigars—just like in cowboy movies—while waging war in earnest. For some reason, it seems that the Americans are exaggerating the numbers of soldiers in every Viet Cong regiment. According to estimates, there are three regiments in the area—the 32nd, 33rd, and 66th—out of a division headquartered on the Cambodian border.

Each Viet Cong regiment must probably muster its own men (obtain reinforcements), and bring in supplies. Perhaps they search for provisions—additional food—on their own initiative), add and oversee—impose their will—on the tax-paying population (French tea plantation owners, those paying other taxes). For this purpose, they must maintain branches all over the country, even in the plains. If their center is in or near Cambodia —if the men who have been captured belong to the above-numbered units—32nd, 33rd, and 66th—the regiment is probably scattered. It does not intend to attack, to ensure transportation paths to the valley, supplies, 'taxes' and reinforcements.

This question is doubly important, to determine the extent of forces standing against the Americans, but also to predict whether the Viet Cong are planning an attack. I do not suppose that the Viet Cong, after being convinced of American power, are still planning an attack. They would have to lack all military logic, to concentrate large forces in the face of such a powerful (American) enemy with air, artillery and ground facilities. By contrast, the Viet Cong have no planes, artillery, or armor.

I don't want to say anything during the briefing, but for some reason General (Norton) asks me forthrightly, in the presence of everyone—if I have any questions. I hint at my theory, ending with a question mark.

6 August 1966

Yesterday, I went out on operation with a battalion of the 3rd Division, seconded to 2nd company, in its first landing in the area. (I make continuous notes).

The procedure is as follows: First, a 20-minute air barrage, after that—artillery. After that—divisional artillery. After that—covering fire with 50 mm machine guns from the landing unit; and then—the landing takes place.

I go out on ambush patrol with regimental reconnaissance. I remain with them (we overnight in one of the gorges) and today—two days and nights. At noon, when the food arrives, I return (with a journalist, and my accompanying cameraman, Stone from United Press). The patrol is both interesting and informative. Interesting—because of the dramatic American activity; informative—because I believe they are on the wrong path.

Conversation with the Chief of Staff
6 August 1966

When the first helicopter (with the battalion commander) arrives, all the journalists run to board it; however, he tells them that he has come to take me only (the commander's machine, because of the heavy wireless equipment, can only take one additional man aboard). I am scheduled for dinner with General Johnson that evening.[23]

I admit to having no idea who he is: I had thought that he was Westmoreland's chief of staff but learn from the battalion commander that he is higher than Westmoreland by several ranks, and the most senior soldier in the army (only the chairman of the joint chiefs of staff ranks higher). I am taken, in my 'Cinderella-like'[24] muddy state, to battalion headquarters, where I find my knapsack: from there, to the brigade commander, named Daniel. He cannot hide his excitement: he himself was not invited to this honored dinner. When they had started looking for me—somewhere in the jungle—they were not pleased (everyone thought about prior jungle encounters, some with heavy losses). He doesn't even allow me to eat lunch with him (not, God forbid, from ill will), but sends me at once with his private helicopter—to division staff. My escort, Captain Brafi, tells me about all the fuss; they were concerned that they might not find and take us out of the jungle in time (they also had to find a suitable landing site). As soon as we arrive, he hurries to inform the 'boss'—that everything is OK.

My problem is simple—how to 'peel myself out' of my muddy encasing. There is no running water, hot or cold; there is nobody to polish my boots, even the kitchen is closed.

First of all, I request 'protektsiya' from a corporal in charge of the senior officers' mess (he already knows me), and receive three hot meals (for the cameraman, captain, and myself). After that I find the room maid. I give her my boots and 100 piasters; she takes the hint. I snitch some soap powder from the kitchen in a piece of newspaper, and water to shave and brush my teeth from a jerry can in the toilet. The cameraman (a dear young man) is quite correct: the body of anyone who walks in the jungle acquires a skin color of a two-day-old corpse.

At 6.30 pm I put on a clean 'suit'—battle fatigues that I washed the day before yesterday (very wrinkled—but clean) and go into dinner.

A group of men stands at the bar: some ordinary officers, others who arrived with Johnson, also the Vietnamese commander of II Corps—in whose district the current operations are being carried out. I am placed at Johnson's left (the host, Colonel Lee, to my left). The Vietnamese general sat facing Johnson, with Norton to the left of Lee

I speak with Johnson during pre-dinner drinks, and at dinner—at both his and my initiative. I start with my own frank and critical assessments of their operations. I don't want—if I put it in writing and he or Norton reply—them to think that I praise them in public but criticize them behind their backs. I tell General Johnson and Major General Norton, the division commander, that I am impressed by their firing and killing power, but less so with their tactics of search (following the slogan 'search and destroy'). To begin with, it is a tactical error for our patrol to keep on walking, instead of stopping while it is still light, determine our position, and set up camp for the night in a suitable location. As it was, we simply stopped when we were completely exhausted, not knowing what was on our right or left. But that was not the most important oversight, and these were decisions made by junior captains. I consider the main error to be in the way the patrol commenced. After our patrol arrived by helicopter, we set out in search of the Viet Cong. Therefore, if there was a Viet Cong unit nearby, they would receive advance notice of our arrival, especially as it was preceded by aerial bombardment. If the Viet Cong decided to escape, they would have done so before our patrol arrived, using their superior knowledge of the local terrain. So, what was the patrol's chance of catching up with them? And how likely could we catch the Viet Cong, since we had left our helicopters behind and were now travelling continuously on foot from the landing site onwards (at most, during half the first day, a distance of 1,500–1600 meters, and after that camp for the night—another issue)? If our patrol wanted to meet up with the Viet Cong, we should have landed in separate locations, surrounding the expected Viet Cong position. Thus, any departing Viet Cong units would be ambushed by the patrol (if roads were suitable). This way, we could have cut off the Viet Cong escape routes rather than land in one place and try to follow them on foot.

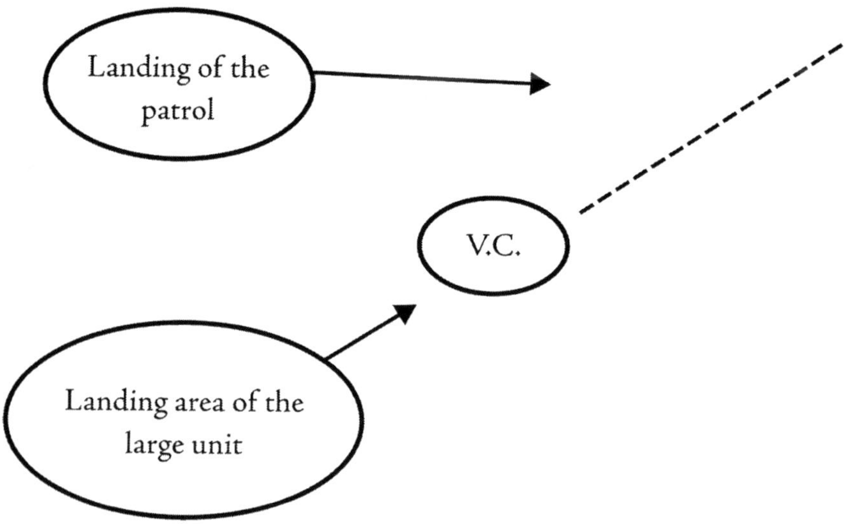

But the Americans are not doing this: they are more concerned about removing their men from their own artillery fire; they worry (they have—as the French before them—had bitter prior experience with friendly fire) about small units of their own men mixed in with large Viet Cong units.

So, the principal question (the key, in my opinion, to America's failures) in the last large operations—first 'Hastings,' and now 'Paul Revere'[25] (the 1st Arborne Division operation which I am currently attending) is that they want to have their cake and eat it: to catch the Viet Cong without putting themselves in too much danger. So as not to put themselves in danger, they operate in large concentrations, with a perimeter protected by artillery on the ground, preceded by bursts of heavy fire. However, all these, in their systematic sequence, simply allow the Viet Cong to escape. In a jungle region such as this—the Viet Cong can either escape or hide.

The other way would be to land—without prior bombing, and as quietly as possible—reconnaissance units, who would establish contact with the Viet Cong. At the moment of contact two things should be done: strengthen the unit making contact, and (according to where the Viet Cong were found) bring in additional units, to surround them, and ambush their paths of retreat.

I do not think that, with the powerful and successful facilities available only to the Americans, the Viet Cong can accept such odds. The Americans have at their disposal all kinds of air assistance (mainly strategic), possibility to lay down rapid artillery fire, airborne artillery, landing (or parachuting) more troops. Although what is being done is called 'search and destroy,' in reality, it is attack, with a defensive psychology, which removes the element of surprise and rapid movement (because it is limited to large troop concentrations). I estimate that, until now, such encounters have been initiated by the Viet Cong themselves, or by American inattention. However, if the Viet Cong concludes (I assume that they already have, or soon will) that they cannot measure up to large American units and want to prevent such large operations—the American way of fighting allows them to do so.

I understand the American's purpose completely—to prevent fighting. Everything must be done to save young lives. Heavy use of artillery and airpower on every sniper is therefore justified. However, when the desire to prevent losses creates a defensive 'ballast' that becomes too heavy—attack turns into defense.

When I tell him that I am more impressed with their strength than their 'search' tactics, General Johnson says: 'No, No, Norton is a very good "searcher."' He freely admits that he has different commanders, with different attitudes and qualities (he suggests that I visit the commander of the 1st Division)—he is aggressive, in contrast to the conservative commander of the 25th. (He adds: 'perhaps I have to say this'). But he does not intend to make them change their approach: he lets each commander do as he sees fit.

In his opinion, the American Army's deficiency (I don't know whether he means Vietnam, or in general)—is lack of sufficient intelligence. They are working with a conqueror's brutality, bombing the entire area, not targets pinpointed by proper intelligence.

In any case, what are the war's objectives? To create conditions and environment for a lawful government to do its job (he obviously means only Vietnam). How can this be achieved? By defeating the enemy. Not destruction, rather defeat. How can they be defeated? By movement and firepower. What is preferable, movement or firepower? Movement, because destruction is not the aim.

His words lead me to believe (I hope correctly) that he is not interested in increasing bombings and use of artillery in Vietnam; he prefers movement of 'cavalry' units—even if they allow the Viet Cong to flee. Destruction is not his aim. (I tell him that, in the middle of the Sinai Campaign, I ordered that as few Egyptian soldiers be killed as possible).

When I ask him whether he is pleased with the progress of the war in Vietnam (I add: perhaps it is not right that I pose such a question), he says: 'Not really. We are enlarging the army here. Perhaps there is no alternative—but it is not a good thing. When we accepted the burden of this war, the Vietnamese Army had (?) 40,000 losses. The Vietnamese had already—relative to the population—suffered as many casualties as American during the Second World War (430,000).' I criticize here for the seventh time—war always has its ups and downs. Johnson maintains that, in March 1965, conditions were bad; now they are much better. Because of what is being done here, there is also development in Indonesia; but this development will be accompanied by serious economic problems.

In some or other context, he asks about *Nachal*[26] (I don't remember which settlement. but I do remember the conversation). I tell him that we have helped establish *Nachal* in other countries as well—Burma and East Africa. I explain the importance of Vietnamese tribal border settlements (from our experiences in Ethiopia and Burma). He does not know Africa and has only passed through the Middle East. I ask whether he would visit—he says he doesn't think so, because he is also occupied with Latin America. When he gets up to leave, he asks whether I would return through the United States; I tell him I don't yet know. He says that, if I do, I should call him, and we could have lunch together. I thank him, and he repeats what he said: 'if you return through Washington, call me.' I promise to do so. After he leaves, General Norton comes up and says to me—half in jest, half seriously—that he is angry with me—I promised not to go on reconnaissance patrol with an infantry company, and I have. I am McNamara's guest, and they have received strict instructions to look after me. Everything else is allowed—just no patrols.

I explain my motivation to him—I am an infantryman, and, if I am not on the ground, I can have no feeling for the area. I want to know the feelings of the Viet Cong when they are being shot at, etc. How one moves around in the jungle. The Americans do this with helicopters—but the Viet Cong do so on foot. I want to know what it is like to walk in the jungle, when at any moment a man could encounter the enemy. I see that he understands; in any event—we part amicably, his mind at rest.

I tell him that I do not keep my promises, but that he certainly does. He tells me that if there would be operations with his units—he would take me there. It would be worth my while to remain there for a few days, so I postpone my trip back to Saigon. What is going on?

He thinks for a moment and says—'perhaps you will not have to leave tomorrow morning.' He writes something in his notebook and says that he would probably attach me to a long-range patrol (the trouble with you, he said, is that we like you too much).

I suspect that I will have to—for the umpteenth time—change my plans. In any event, I will not leave the 'cavalry' while it is in intensive deployment.

Before my meeting with Norton, I had arranged with the director of civil operations to visit the area with him tomorrow and the day after. I had arranged my trip back to Saigon for Tuesday 9 August 1966. Now I must cancel both. I want to see civil operations here as well; we'll do that in the morning.

Meanwhile, when I returned towards evening, I find a message from the Saigon press attaché's office:

1. General Johnson invites me to dinner tomorrow evening. All sorts of 'important people' will also be present.
2. Westmoreland and others are prepared to meet, whenever convenient for me.
3. The Chinese ambassador wants to arrange a dinner in my honor.
4. They have begun preparations starting the day after tomorrow for my visits to civil operations.

I had asked to be excused and postpone everything until Tuesday; now I shall have to do so (postpone) again. Likewise, to escape, courteously and so that no one loses face, from the Chinese dinner (from Formosa, of course). That's all I need!

Unless additional changes occur, my program will be as follows: Operation 'Paul Revere' tomorrow and the day after (Sunday and Monday); civil activities—Tuesday and Wednesday; travel to Saigon on Thursday 11 August. At night I bathe properly and do part of my laundry. I have no idea when I will get rid of the remaining dirt.

7 August 1966

I meet Johnson again at breakfast—he is stuck there because of bad weather. Without being asked, he turns to me and shares what is on his mind: the delicate balance between staying put and continuous movement. The Marines, he says, are inclined to stay in one place and wait for the Viet Cong to return and attack them. However, the 'cavalry' wants to move. The question: which is more correct?

I tell him (as courteously as I could) that changes in Viet Cong tactics must also be considered. They will not repeat what they did a few months ago. They have learned a lesson. I ask him whether he thinks the Viet Cong are still planning to attack and destroy large American units. He confidently replies—yes. Yes, whenever we land somewhere, they move their troops towards us.

I tell him I have my doubts about this. It may be that, at first, they planned to do so (Ho's 'three stages'), based upon their experiences with French and Vietnamese units. However, in the meanwhile they have learned a costly lesson from American air power and artillery. They will not try to do so in future, but rather look for contact with small units, or adopt 'hit and run' tactics. It is also likely that their units are much more split up than was previously thought.

They catch a prisoner from the 33rd Viet Cong Regiment and assume from this where the unit is. However, the unit is probably split up: part in the plains—to obtain supplies but, mainly, to maintain their influence on the population—part between the plains and mountains, and part in the mountains.

We do not continue. His initial words—that the purpose of the Viet Cong is to attack and destroy large American units—surprise me. I gain the definite impression that this idea—the heritage of every senior American commander—leads to a policy of defense, in camps as well as mobile units. As I saw yesterday, this policy seems mistaken.

This does not mean that the Viet Cong might not attack American units with concentrated mortar fire—or even, during calm weather, surround and destroy small, unprotected American units. However, it should be assumed that they have learned—or are learning—their lesson: What succeeded with the French will not work with the Americans. If the Americans do not change their viewpoint—their tactic is bound to have disappointing results: a lot of hard work, with little to show for it.

After Johnson leaves, I continue the discussion with several officers. They are less optimistic than the person who told me (Barnes [*sic*] or Norton) that the Americans can operate during 90% of the monsoon season. Correct, it is possible each day to fly from one place to another, but they have no confidence that they could arrive at any given place from another.

The 'Cavalry'—Continued

I am still with the 'cavalry.' At 1.35 pm, meeting with General Jack Norton.

At 9.00 am the next morning I go out with Joseph Hoal,[27] deputy director for II sector (corps) under Vietnamese General Vinh Loc[28] (the country is divided into four such sectors). Hoal oversees American aid. All other American agencies dealing with Vietnam are subordinate to the director of welfare service. There are forty-four provinces in Vietnam, in four districts. Pleiku is the largest and most neglected province. The administration does not want to come from the capital to work here: it is far from the rich coastal region.

The director of welfare services deals with everything: health, education, agriculture, building, etc. There are other foreigners here: Swiss, Koreans, British, etc.; also, American students, a total of 250–300 foreign students working in these areas.

Hoal tells me that the population are interested, first and foremost, in education and health.

Although a man of few words, he makes no secret of what he sees as the principal problem: the discrepancy between the military presence here, and its consequences. They wish to establish local administration, and influence hearts and minds. But the very existence of the army has the opposite effect. Americans pay high prices for products, and the Vietnamese cannot compete. They receive the best service (in restaurants, hotels, etc.). There is also the issue of harassment of women—Hoal's secretary is Vietnamese, and full of complaints about how his young wife is treated. It is impossible to make the cities 'out of bounds' for soldiers. In war, there is bombing and destruction, and civilians fleeing the fighting area become refugees. There is no lack of money, but the Vietnamese administration (including welfare) is incapable of properly allocating the budget.

The main problem remains the discrepancy between military—search and destroy—and civilian—hearts and minds—slogans. (Perhaps to this should be added General Johnson's maxim—to defeat the enemy without destroying them).

In the meanwhile, the telephone rings: I must be at the camp, to depart by air with a reconnaissance unit of scouts at a distance.

We depart for the unit (squadron) but land in the wrong place. Finally, we find them (I have no idea how anyone can find anything amongst these hundreds of helicopters and thousands of tents scattered all over the area).

The unit commander explains the battalion structure to me (he gives me the unit manual).—This is a divisional battalion tasked to reconnoiter, by foot, ground vehicle or helicopters; they are responsible for deep reconnaissance. Their current mission is twofold: 1) Opening (existing) roads; 2) regional reconnaissance.

As far as 'opening the roads' is concerned, it turns out that these roads are travelled only by convoy—In fact, even after they have been 'cleared and opened' the convoy system must continue. Their task is, therefore, 'demonstration of power.' They travel on the roads—by convoy, etc.—and reconnoiter parallel to the roads in the surrounding area. His words about reconnaissance astound me, because this is exactly what I thought should be done. According to his words, the unit is divided into small groups of six to eight men, who reconnoiter the area to find the enemy. Their task is not to fight, but rather to be the division's eyes and ears. When I ask, he replies: if they encounter the enemy, they split up further into pairs and return separately to camp. They can then receive firepower assistance from:

1. artillery (in other words from camps scattered in the area—from which almost all locations could be reached);
2. airborne artillery units of the 'cavalry';

3. the unit itself. In exceptional circumstances, if they were outside the range of all help (very bad weather conditions), they would not go out at all.

According to the commander, air reconnaissance at tree-top level is the most useful: no one can hide from it. His words exude self-confidence. He tells me that his spotting methods have revealed Viet Cong units that they themselves had not found. Every time they meet up with the Viet Cong (large units—battalions—four times during February), the Viet Cong have wanted to prevent the encounter—this is exactly the opposite of what I hear from other units. Only when the Viet Cong catch disorganized American units—at the moment of landing in the area, and especially in the evening—do they dare attack.

Since February, his units have killed about 500 Viet Cong; by contrast they have 50 dead and 100 wounded. Their task is not to kill, but search for and find the enemy.

Even if he drops a significant portion of men in the area, the unit seems so filled with confidence, that they would allow small groups to go out into the jungle (in the full knowledge that aid would come).

In the meanwhile, news comes that their (ground) patrol has encountered the Viet Cong, and that they have two wounded. We stop talking and go to the helicopter. They dress me in a bullet-proof vest (I tell them I feel like a Chicago policeman), and I sit to the pilot's left. They show me how to operate the machine gun—just in case. I say that it would be better if the American Army did not depend on me—I have not even fired a revolver for 10 years (and in any case, I have come to observe, not fight). I would (of course) fight, but only if it was necessary.

The pilot's enthusiasm surprises me. During many hours of very low flying (really at tree-level) I have not distinguished a human form. The pilot tells me that the Viet Cong disguise themselves amazingly well and hide under the treetops. If they don't open fire—it is impossible to reveal them from the air. We fly for 4 full hours with four helicopters (filling up with gasoline, and returning). There are clearly Viet Cong in the area: one of our helicopters (not the one in which I flew) was hit by a bullet; the two men on patrol have been wounded in this area a short time ago. Even though I track and search—I see absolutely no one. The pilot says that he has seen several shapes, and the helicopters open fire on the approximate area. (The spotter pilot drops a smoke grenade—white or red—to mark the place for others and for himself, since he might lose sight of it while making more passes over the target.).

In sum, helicopter fire is very scattered. But it is so heavy that it has to hit someone: there are probably several Viet Cong on the roads at which the pilots shoot, but I cannot overcome the feeling of lack of proportion: four helicopters firing (each with three machine guns and rockets) on a few Viet Cong hiding in the jungle. But this is the reality of war here.

I cannot help being impressed by the proficiency of these operations. One of the helicopters lands next to the foot patrol, who gives them details on the enemy's direction. The helicopters are mainly the 8–13, the 'scouts': the cabin is completely transparent—exactly like hunting dogs, twisting around in the wadis, descending between the trees. The helicopters have neither doors nor windows; everything has been removed, to increase the area of visibility.

Refueling, with plastic tanks that look like large barrels is done at a field station and only takes a few minutes; they don't even turn off the motor.

The nearest helicopter informs us that they have killed an elephant (it is said that the Viet Cong use elephants for transportation). At the end of the tour, their assumption is that they have hit several Viet Cong. That may well have been the case, but I did not perceive a single image during that trip.

When we return to base, a young pilot named Cohen comes up to me to say 'Shalom.' A soldier named Perlman takes us back to division headquarters: his parents are currently on a visit to Israel. He is waiting for his year of service in this damned war to end, so that he can return to his family. He asks that we be photographed together and doesn't hide his disgust for this entire business (the war here)—but he doesn't know of any better way.

I have no doubt that these reconnaissance units have real power to find the Viet Cong, by air and ground observation. But I assume that the Viet Cong will reach some sort of co-existence with this as well. Every unit has its limits: helicopters and planes are noisy (at first, they come as a surprise at low altitude—but they circle around the area for a long time). They cannot function at night and are also strangers to the area. Obviously, facilities for communication, movement, observation (from the air, while the Viet Cong have observation ability from the ground only), enormous firepower—are all to the Americans' advantage. But for all that, I do not leave with the feeling that the Americans would 'checkmate the Viet Cong in three moves.' Obviously, the Viet Cong cannot seriously hurt the Americans: but they certainly could change to, and maintain, guerilla warfare. Through the entire time, the feeling of 'co-existence' will not leave me: as if Vietnam is a sort of a sack containing two kinds of potatoes. The see-sawing and turmoil with which this war shakes the area will, at the end of the day, make the large potato (America) 'get along,' side by side, with the small potato (Viet Cong)—which will find a place in the nooks and crannies between the large ones. In the mountains and jungles of this country, amid the local population on the other side of the Cambodian and North Vietnamese borders—there are abundant nooks and crannies, into which the Viet Cong can crawl, and remain alive and kicking.

During dinner, Colonel Lee (advisor to the Vietnamese general) tells me that Johnson's request that I visit him in Washington and meet him for lunch is serious—not just a courtesy. He compliments Johnson highly: he is the best of the last four chiefs of the general staff (Ridgeway, Taylor, Lemnitzer and Johnson[29]). He is a moral, very religious man, and his words about the need to 'defeat, not kill' flow from a deep feeling of humanity. He asks for an hour of my time before I depart.

At 11.00 pm. I am sitting barefoot in pajamas, all the washing that I had done that evening (finally hot water arrived) hanging from the windows—when suddenly General Norton arrives. He has been in conference with his subordinates, and tapped them on the shoulder, complimenting them on their efforts. He sees a light in my room and comes in to greet me. I give him a copy of my book with written dedication. He says that he has prepared a plan for me—tomorrow—with Barnes [*sic*]. I tell him that I had promised to be with Hoal the day after tomorrow and Wednesday. He smiles: 'who knows whether you might remain with Barnes [*sic*] for a few days? Who knows what will develop? Listen to me and remain with this division: you won't be sorry; stay with us.'

I would not be surprised if I must change my plans yet again. Today I inform Hoal that I will remain with him the day after tomorrow; he has prepared a tour of the surroundings for Wednesday. They have already informed Saigon that I will now arrive on Thursday (this is the umpteenth change already). In any event, while I continue seeing interesting things here—as has been the case each day so far—I will stay here, even if it causes inconvenience for myself and others. This is exactly why I came here from Israel—to see how the Americans are waging war.

Norton tells me in a paternal tone that it would be better if I rested, because difficult, tiring days lie ahead. In the meanwhile, it is already midnight.

With Units of the First Cavalry Division

This morning, at breakfast—a lengthy discussion, initiated by Colonel Lee starting with the question: how to defend this base from a large-scale Viet Cong attack. I ask whether he still thinks that such an attack will occur. He replies—certainly, yes. They have convincing reports that the Viet Cong have not yet been weaned from their perception of Ho's 'phase three' of war. I say that I do not believe it: surely, they have learned a lesson and now understand the futility of such action. It is interesting how the American psyche still fears an offensive. I ask when the last attacks had taken place. It turns out—in March; and even then, they were near the Cambodian border (the Viet Cong thought they could hit the

Americans there and then escape across the border); or in the welfare administrator's camp (most of the soldiers there were not Americans); or in the west—which presented a unique opportunity. Lee does not change his opinion, and we make a bet. In any event, if they don't attack by the end of the monsoon—the Viet Cong will have wasted an opportunity. The monsoon starts later in the coastal plain.

We speak about the official American position on the war. I say (those present nod in agreement) that, the longer this war continues, the less popular Americans will become: the war will only be supported by a clique of generals and colonels. The country is being deprived of its true (not formal) independence. The people are not able to compete with American standards and are being relegated to 'second class' status, at a time when they are in their own eyes, without Americans, 'first class.' He gives me the example of Korea. There the 'population' don't want the American Army to leave. I, in turn, give the example of Burma: they believe that they are historically lucky not to have requested American aid. Then we move on to Europe. All countries, except France, want the Americans to stay. I tell him that this may well be, and I am no expert. But in Europe conditions are different—there is no cult of personal supremacy, like here. The Europeans do not stand idly by, just because the American Army is there. By contrast, here:

1. The population is inferior in rank and status.
2. There is a large military presence.
3. The population hate the war.
4. There is an American administration that acts like a custodian.

All this is excellent material for Ho Chi Minh, to preach to the youth that the Americans are foreign conquerors, who must be opposed.

In any event, I believe that Ho Chi Minh is likely—if the war continues—to be forced to change his method of waging war to guerilla tactics, with units across the northern and western boundaries.

The main fact that comes out of our morning conversation, is that the Americans still believe that the Viet Cong will attack in strength. This is the perception even after the suffering of Operation 'Hastings': 'the monsoon attack error.'

In the morning, we go out to the second brigade of Colonel Berenzweig (for some reason, I have called him Barnes).[30] The trip is postponed because of bad weather, with touchdown now scheduled for 1.00 pm. We walk around the area a bit—I visit a primitive market where mountain people sell vegetables, fruit and fish. Mud and rain everywhere. Girl-women breast-feeding babies hanging from their backs.

We leave at 1.00 pm. An extremely impressive image of dozens of helicopters swooping into the jungle like a swarm of wasps. Flight at absolute tree-height, because of low clouds. I fly with the support helicopter (each company lands on their own spot—except for the battalion staff helicopter). No one knows where they might encounter the Viet Cong. There is no front line, no 'in front of the lines,' no 'behind the lines.' There is only jungle, and the enemy may be anywhere or everywhere.

The battalion commander suggests that, if the first company helicopters are fired at on the ground and it becomes 'hot,' I cannot be landed there, but go on to the 'landing zone.' I refuse and say so to Brigade Commander Berenzweig. He chews on his cigar and says: 'As you wish. In any case, we don't know where the danger really is.'

When we approach the landing area, we see planes returning from preparatory bombing of landing areas, from which smoke—caused by airborne bombing and ground artillery—is rising. We land in one of the largest forest clearings I have seen here. Even the grass is shorter—the height of wheat in the Jezreel Valley (in a good year). The area looks as if it has been prepared for the planting of tea, coffee or rubber trees.

We do not encounter fire. The sun warms us and the sky clears. There is an almost festive Pentecostal Shavuoth[31] atmosphere in the air, to which newspaper and television cameramen are added. But, instead of the first fruits of Pentecost, large helicopters land, one after the other, and lower 105 mm cannon, howitzers, ammunition, jeeps, etc. These are followed by a crane [CH-47 Chinook or CH-54A Skycrane] carrying the 155 mm, a truly terrible beast! (Cost—$3,000,000.00!).

The artillery takes no time at all to set up, and immediately opens up preparatory fire for a second landing site, which isn't within the range of existing artillery in the area (the field in which we land is too far for artillery preparation).

In the meanwhile, we receive news that another company (originally planned to be landed first) is being fired on, and that a battle has developed. We decide to go there and look for a 'ride.' We are given place in a helicopter returning to base; we return with it in the hope of flying in with the reinforcements for the company which has been attacked. But complications develop and, by 5.00 pm, we still have not left the labyrinth of camps, so we achieve nothing. I have already despaired and want to return to the base by jeep, but at the last minute we find a helicopter from the 3rd brigade. We fly with them, and I find the brigade commander—Colonel Daniel[32]—in the same helicopter. I explain what I want to him. He says to me: 'if you get wounded, my neck will be in the noose'—he points to his neck (a thin neck, unable to carry a thick rope). I reminded him of what Churchill said: 'some chicken! some neck!' (his 30.12.1941 speech to the Canadian parliament about the prophecy that the British would have their necks wrung like a chicken by the Nazis). He laughs, and gives me his own personal helicopter, on condition that he drops us off at battalion headquarters of the fighting company (from the second brigade, which doesn't belong to him, and in whose affairs he cannot interfere). This proves to our benefit. We fly to battalion headquarters and there—in the area of the very small landing—a surprise awaits us. Next to the guns, hunched over maps of the area spread out on the ground, are General Norton, General Walker (from the 25th division, whose brigade was responsible for this area[33]), Colonel Berenzweig,[34] and their staff officers. They are planning reinforcements for the unit that has been attacked, to block and encircle the Viet Cong (Norton nods to me from afar). They estimate the presence of three battalions of enemy forces (each with 300–350 men). The company had not been attacked on landing, but rather after they advanced in the direction of the stream—about 1,000 meters from the landing area. The Viet Cong unit had entrenched themselves there: and when the American company advanced on the road, it found itself attacked from three sides. At this stage, they had about fifty casualties (killed and wounded) but information was sketchy. It was not clear whether the Viet Cong had opened fire simply because the company had come upon their positions and forced their hand– or whether it was a planned ambush. (In this area there are main roads leading to Cambodia, and American patrols may have passed along them—at the most platoon strength—and reached the river crossing, where the clash occurred). If the Viet Cong had not been forced, they probably wouldn't have attacked, at least at this exact time, but rather have prepared a night ambush.

After having escaped from their planning session, they come up to me. Barnes winks: 'You got onto the wrong plane.' Norton says that the area should be evacuated to allow the regimental commander to pass through. I tell him that I have come on Daniel's plane, and that he has returned to base. Norton invites me to join him.

Norton flies the helicopter himself; communications are suitable for command, and the inside is also a little fancier (upholstery). His assistants and the pilot fly with him. When we get off the helicopter he says to me: 'join us for dinner: I want you to learn more.' I have no idea where we are; as it turns out, we have arrived at General Walker's headquarters (3rd brigade of the 25th [Infantry] Division). I have not met him officially yet, and he introduces himself. While Norton is still busy, Walker explains the details of the encounter (the area lies within his control sector). His forces have already had one skirmish, a week ago, in this exact area: a platoon was ambushed in the same way. He had called for urgent reinforcements and sent in additional firepower; seventy Viet Cong were killed, with probably double the number of wounded; in other words, an entire battalion of casualties. After that—4 days ago—they revisited the area and found it empty. But now the Viet Cong have returned. They probably have some sort of depot in the area (rice, ammunition– a large quantity of explosives was found in the bunkers during last week's clash). Perhaps they see a chance of an American patrol passing by. This area—about 20 kilometers from the Cambodian border—is a refuge for retreating Viet Cong units. Walker illustrates his words with diagrams on the tablecloth. Meanwhile, Norton arrives. We drink beer and eat—a meal in the field, but it is good.

Walker gives the impression of being more reserved than Norton. When I say that the company with which I landed is not within artillery range of another unit—he says that such a mistake must not be made again. Each unit must be sure of covering artillery fire. He completely lacks the ***èlan*** of a 'cavalry' commander.

After lunch, Norton returns me to divisional headquarters (by plane). On the way back, he tells me (while piloting the plane) that he placed my book on his desk last night; three officers have already asked to borrow it, but he refuses until he has read it himself. When we part, I ask what I should do to reach the battlefield tomorrow. I want to see the positions, etc.—even if contact is broken during the night. He says that he will send his assistant to take me there, and I tell him not to forget. He says angrily—'I don't forget you, but the last time I came to take you, I found that you had run away!' (He still has not forgotten my going out on patrol). Afterwards he smiles and winks: 'be ready in the morning!'

During our conversation—on the plane and at lunch—Norton repeatedly emphasizes the need to 'be quick on the trigger.' He criticizes commanders who do not land their men rapidly near the scene of a clash. The enemy should be confined to a limit of 2 hours (until darkness falls) and destroyed. Two American character traits stand out in the 1st Cavalry Division: 1) first-class efficiency; 2) unique aggressiveness.

Anyone who has seen the groups of commanders hunched over their maps—and afterwards Norton jumping into the helicopter and piloting it himself, precisely at tree-top level—must lack all imagination not to see the analogy with the heroes of the 'wild west' and 'cavalry' in the movies. When Norton says 'to be quick on the trigger,' while he, Barnes [sic] and others are chomping on cigars, you see before you images of their cowboy ancestors who were 'quick on the trigger.' However, this time, instead of jumping onto the saddle, he jumps into a helicopter and flies it himself. Especially here, with the jungle background, it seems like a fight between sheriff and Indians.

But issues are too serious for joking. When they use orthodox weapons, the 'cavalry' is a fearsome force—rapid and efficient. This division operates its weapons—including tactical and strategic artillery and air assistance—with first-class efficiency. I assume that those occupied with military matters are following the operations of this division with eyes wide open and will not be surprised (without any direct comparison) like they were by the crushing force of the German army during the Second World War.[35]

I have not changed my mind—the Viet Cong have no chance at all to defeat the Americans, by attacking with large divisions. I am assuming that the Viet Cong have learned their lesson, and that today's encounter occurred by accident, not on purpose. If they have not learned by now, they certainly will soon. This time, they will certainly have very heavy losses. Any attempt to move on jungle paths at night, burdened with mortars, wounded, jungle equipment, when American airpower is at its strongest and artillery cannons (with unlimited ammunition) are firing on them—would be catastrophic. I assume that the only thing the Viet Cong can do is split up into even smaller units, and scatter in different directions.

I read today in the local American military newspaper, that, according to General Norton, Operation Paul Revere is meant to disrupt Viet Cong preparations for an attack during the monsoon season.

This declaration is (with all possible humility), totally stupid. The operation is being organized because the 'cavalry' is located here and would like to fight and pursue the Viet Cong with all their strength, day and night, monsoon or no monsoon up to (and perhaps a little beyond) the Cambodian border. The division is built for attack and mobility; any time they assume that Viet Cong units are somewhere in the jungle, they will push their planes from the landing strip to landing strip, organize air and land patrols, and rain fire from the air, and fire and artillery from the ground, on anything that moves—all with no connection with the 'monsoon attack.' The 'cavalry'—is an engine of war, and its commanders are thirsty for war, or —to be more exact—to use the machines that are available to them. Here, they have a chance to do so, against a designated target.

General Norton drops me off at the camp and asks if I can walk to my room. Obviously, I say yes. When I reach the officers' gate, the military police don't let me pass, and they refer me to the mess employee entrance. I am not surprised. I have not looked at how I am dressed—in 'all my splendor'—in the mirror, but I can imagine my wretched state, with a Vietnamese hat (bought in the black market): after rain and many washings, its visor hanging loose, muddy boots, and work clothes without any signs of rank, unit, or army. I have never been the world's snappiest dresser, but now I look—in these clothes, after a day of rolling around in jungle mud—like a kitchen hand in a private's mess.

Despite being so tired, I wash the mud off my boots, otherwise I will not be able to enter the mess tomorrow. It takes me a full hour to get them clean.

When I get to my writing, the subject will not be why the Americans are fighting here, but how they are doing it. Obviously, achievable objectives are an essential part of war planning and evaluation. It isn't too far from the truth to believe that the United States will conquer the whole of Vietnam (it is not far from doing so already), and see, in that, 'pacification' of the entire region. However, in the Viet Cong's eyes, Americans will remain foreign conquerors: they will indoctrinate the youth and fight a guerilla war with the (American) conquering army. General Ky[36] and his ilk will appear to the Viet Cong like Quislings[37] collaborating with, and profiting from, a foreign government. I think that the next stage in American operations will be—opening the main roads for general use.

If the Americans do this—despite its positive effect in principle—it will require roadside patrols and security checks, camps and convoys. All this will require extension of the American Army's reach—I do not know how possible it will be to conscript more Vietnamese soldiers.

Conscription of the mountain people as 'special forces' will, first and foremost, be based on financial compensation. This is also a useful way to prevent them from joining the Viet Cong, binding them to the government.

9 August 1966

Today, I go out into the field at 1.00 pm—before then, the weather is so rainy that it is impossible to fly. Yesterday's news of the fighting—up to now they have counted about 100 Viet Cong bodies (there are certainly more scattered over the area). But American losses are not light—twenty-five dead and about fifty wounded out of the 130-man company. During the night, the Viet Cong break off contact: by now, they have surely split into small units and scattered. They bring me to the divisional reconnaissance battalion, located on the Cambodian border (I visited them once before). When I arrive in pouring rain, the deputy commander, Major Shattuck—a pleasant young man—is sitting under a dubious-looking poncho, speaking vigorously into a radio.

He explains to me that a group of Viet Cong has been found, and that an infantry unit has been sent to investigate. He suggests that I join the plane going to the area, and I agree. During the flight, it becomes clear from radio communications and the pilot's explanations that the unit (company) commander, Major Belkley,[38] has left that morning in a small two-seater helicopter 8–13 ('Bell') with a transparent cabin, and found the enemy unit. He and another plane have been circling around the Viet Cong for about 45 minutes, pinning them down with machine guns. A fifteen-man infantry artillery unit is about to land. Radio transmissions are very clear, and I hear the commander in the air directing the unit—which has meanwhile landed—to advance towards the enemy, warning them to avoid ambush. He really leads them, like someone holding a child's hand: 'Halt! You are 100 meters from the Viet Cong! Be careful; two of them are lying between the bushes on the right-hand side; go back a little and divide into two groups,' etc.

Within an hour, the entire group of eight Viet Cong has been caught, and transported to base by helicopter. We return there as well, to observe the prisoners.

Commander [Major] Belkley tells me that he has almost run out of gasoline but understands that it is within his power to pin down the unit. I understand the reason later. The Viet Cong have a wounded man, and only one rifle between them: the remainder transport mortar ammunition. They could not fight in such a state. Were it not for that, they would have opened heavy fire on him and tried to escape or fight.

This American unit has an adventurous spirit of discovery.

Afterwards, I fly with one of them, Captain Carmichael, in his OH-13 for an additional tour. The plane is open on two sides and pounded by rain and wind. The pilot holds the controls with one hand and operates the machine gun with the other. The second observer (myself) is given an American automatic rifle: not attached to anything, simply operated from the shoulder. There is no door, and the pilots wear bullet-proof vests. The pilot flies slowly—18 knots an hour, and very low, almost between the trees.

These Americans have no ground artillery cover, neither are they 'doing big business' with the Viet Cong and the jungle (also not from the Cambodian border … .) They don't believe they will clash with the Viet Cong when they land in force, so they will manage as they are. In their general attitude, organization and rudimentary way of doing things, these Americans remind me of our 89th Battalion during our War of Independence. The differences, of course, are that they have basic supplies—with planes, communications, ammunition, and the ability to obtain—in an emergency—rapid air assistance. All, obviously,—at American standards.

There is an amazing lack of bureaucracy here. Planes and men are sent unhesitatingly to where they are needed. The main thing is—to work, and hit the Viet Cong. If they catch someone in the air, on the ground, route L or M, if possible, the call is answered, and the person is moved. There is a pile of ponchos, water tanks, first aid supplies, on the floor of the plane, but mainly crates of ammunition. They fire immediately from ammunition in the wooden crates on the floor. Machine guns, suspended from rubber ropes in this unit's larger helicopters, are shot from shoulder height, during flight. This unit is obviously experienced in jungle warfare, self-confident, contemptuous of insignificant creature comforts or special technology—All this in pouring monsoon rain which soaks and alternatively dries them out every few hours. Clouds come in heavy waves, 'close up' the skies—and then the windows open again.

Prisoners

Six of the eight prisoners are dressed in greenish khaki, with a wide brimmed hat of the same color and age. The other two are dressed in black pajamas over their uniforms. All are North Vietnamese, from a regular unit—the 330th Regiment. They were on their way from the Cambodian border to inside South Vietnam; they were a portion of a large unit that and split up and had to transport the wounded man and the ammunition. The second part of the platoon escaped.

They are not interrogated in-depth but transported by plane to a base in the rear. They have difficulty understanding whatever questions are put to them. The tumult of falling into captivity and everything around them—has thrown them into shock.

If I could have spoken to them and they had been prepared to speak to me, I would have, first and foremost, posed one question: are they volunteers, or have they been forced into service? When questioned what they were doing in the forest, they all answer (apparently as hammered into them) that they have come to fight to free Vietnam from foreign aggressors. Nevertheless, I would very much like to know whether they are young idealists, who know that the road is very long,—perhaps with no return. The fact they had walked for several months to reach the battlefield and are prepared to fight with the primitive weapons and other means at their disposal against the sophisticated well-supplied and -armed American Army, means that either they either didn't know what was in front of them, or were conscripted by force or pressure, simply arriving where they arrived. Obviously, I can't talk to them, but their appearance touches my heart. Young men clad in sandals (in the mud and rain), a large knapsack on their backs, inside a small enamel bowl, an aluminum spoon hammered from a can, a water canteen, an aluminum box for cooking (some still contain a dish of pulses), a small bag of mixed rice and pulses (a full bag comprises the uniform war ration), a first aid package, and personal items such as a cigarette lighter, ink powder (which has first to be dissolved in water), and one of them (? the commander) has a copper badge with a relief of Ho Chi Minh. The load—ammunition, mortar shells, and now the wounded man—is transported hanging from a bamboo stick by prongs or tied to their back with thick rubber straps. The knapsacks contain many straps to bind packages. And they must drag everything—everything—on their backs from Hanoi to Saigon. Their weapons—a rifle of Chinese-Russian origin (writing in both languages) and primitive hand grenades with wooden pins.

The soldiers themselves don't look bad. They are not thin or dirty. Their hair is cut and they have shaved. But they look forlorn and wretched. This impression obviously comes from their impossible comparison with the American Army. A tall black soldier performs a body search to see whether they have hidden papers which could be of value: the Vietnamese soldiers are not even half his height. However, he handles them gently, without pushing or shoving. The soldiers who bring them in from the field treat them the same way—they don't even tie their hands.

I know that external appearances are not indicative of the true face of issues. It is possible to fight a ruthless, active war with primitive methods.

The individual can be an idealist, and sacrifice himself, even in the richest and most advanced army; by contrast, he may be a shirker, hiding behind the backs of his comrades, in a wretched, forlorn army. But a direct comparison between these two armies—in the open field, and jungle—is inescapable, and hits one in the face. The American Army—helicopters, air force, armor, electronic communications, artillery and unlimited amounts of everything—unlimited ammunition, gasoline, spare parts and supplies. Against them—these eight soldiers, who have travelled on foot for four months from North to South Vietnam, carrying shells on their backs, with a little ground rice and spoons made from aluminum cans in their knapsacks.

From a military point of view, it is total insanity for the North Vietnamese to try and fight the Americans. No means of man will help here. The Americans have enough gasoline and munition to pursue the entire North Vietnamese Army, and rain thousands of bullets and bombs on them until they are all killed or defeated.

From a human point of view, this is one of the saddest things I have seen in my life—not just when they sit there, captured (every prisoner, wounded or defeated man is a depressing picture), but also in their 'normal state' in the forest, in uniform, load on back—in the army. The human part of the military is individual readiness to go to war. Two armies fight each other. The Americans fly around the enemy with planes, seek them out in the jungle paths like powerful hunting dogs, and strike, strike, strike without interruption. The North Vietnamese continue to stream into the South, via Cambodia and through jungle paths, so as not to return and be wounded or killed—sooner or later.

Dinner conversation

1. The prisoners I saw today. Everyone around the table knows the picture and agrees that it arouses compassion: individually and collectively (for the entire army fighting the Americans).

2. Another army is being brought to Vietnam (the 4th [Infantry] Division,4,000 men), and 100,000 more. The country is being Americanized. They will solve one (military) problem, and defeat the enemy, with a large army. However, in the process, a second problem will be created—control of Vietnam by the American Army with the mere power of its presence, backed up by civilian assistants. Each criticism is met by the example of Korea, where the American experience succeeded:

   a. The Koreans don't want the Americans to leave.

   b. They have made a quantum leap from the Middle Ages to a modern state—all because of American attention and guidance.

Maybe I do not know Korea. (Who are the 'Koreans?' The 10% of city-dwellers, or the millions who live in the rest of the country?[39] What do these know? How has their way of life changed?). In any event, I receive a clear answer to my question whether America was here out of self-interest—yes. At least they are not hypocrites about it. If the Americans are here out of direct self-interest—that is another, very complicated, issue.

But, no matter what, I must admit America's innate advantage, because it is in their nature to be more active than other nations: they work, produce, advance, penetrate (or try to penetrate) into new areas, and—when they do so—results are apparent. Others—visit, withdraw (Britain, France) or remain, but make no progress (Soviet Union).

One Vietnamese bureaucrat (district governor) has been heard to ask this question: 'but how can I arrange elections, when no one tells me whom they want elected?'

Endnotes

1 Lewis W. Walt (1913–1989), United States four-star general who served in World War II, the Korean War, and the Vietnam War. In Vietnam, he was Commanding General of the III Marine Amphibious Force and 3rd Marine Division.

2 Major General Wood B. Kyle was a notable figure in the Vietnam War. Highly decorated in World War II, he served as Commanding General of the 3rd Marine Division following Major General Lewis W. Walt.

3 Brigadier General Lowell E. English was assistant division commander of the 3rd Marine Division under the leadership of General Lewis W. Walt

4 Brigadier General William R. Fields was a U.S. Army officer who served during the Vietnam War. One of his notable roles was serving as the Assistant Division Commander for the 101st Airborne Division.

5 John Snyder was a notable American cameraman and filmmaker who gained prominence for his work during the Vietnam War. He is particularly known for his documentary work, which provided a raw and unfiltered view of the war from the perspective of those on the ground. Snyder's most significant contribution is the film *The Vietnam War: A Film by John Snyder,* which offers a powerful and sometimes controversial look at the realities of the conflict. His work has been recognized for its gritty realism and for bringing a personal and often critical perspective to the portrayal of the Vietnam War

6 Moshe Perlman (1911–1986), Israeli writer.

7 Charles C. Krulak is a retired United States Marine Corps four-star general who served as the 31st Commandant of the Marine Corps from July 1, 1995, to June 30, 1999. See General Krulak's foreword, this volume.

8 See Chapter 2 note 6.

9 Brigadier General William C. Platt served as the Chief of Staff of the III Marine Amphibious Force under General Lewis Walt. General Platt supported General Walt in overseeing Marine operations and strategic planning in the region.

10 Translator's comment. Dayan could not know how high the rate of post-traumatic stress syndrome amongst returning Vietnam War veterans would prove to be.

11 Chapter 2 note 12.

12 John M. Wright (1916–2014), United States Army officer who served in World War II, the Korean War, and the Vietnam War.

13 See Chapter 1, note 10.

14 General Henry E. Emerson (1925–2015) gained recognition during the Vietnam War for his tactical ability on the battlefield. His tactics as a commander were novel. He conceived aerial reconnaissance and combat methods that were employed effectively against the Viet Cong.

15 Operation Crazy Horse (16 May to 5 June 1966) was a search and destroy mission during the Vietnam War conducted by military forces of the United States, South Vietnam, and the Republic of South Korea in two valleys in Bình Định Province of South Vietnam.

16 This was the Battle of Mang Yang Pass, also known as the Battle of An Khê, from 24 to 29 June 1954.

17 Moshe Dayan, *Diary of the Sinai Campaign* (New York: Harper Collins, 1966).

18 Major General John Norton (1918–2004), commander of the United States Support Command in Vietnam in 1965, Deputy Commanding General of the United States Army Vietnam, and commander of the 1st Air Cavalry Division from May 1966 to April 1967.

19 The 1st Air Cavalry Division, known for its use of helicopters in combat, was a key part of the U.S. military strategy in Vietnam, providing rapid deployment and support in a challenging and often volatile environment.

20 See Chapter 1 note 50.

21 The correct name is Marvin J. Berenzweig, which Dayan subsequently corrects from Barnes. He likely misheard the name at first. Berenzweig was with 1st Cavalry Division.

22 The Geneva Agreements of 1954 established a ceasefire in Vietnam and temporarily divided the country at the seventeenth parallel, with a 3-mile-wide demilitarized zone between the two.

23 Harold K. Johnson (1912–1983), Chief of Staff of the United States Army during the Vietnam War, serving from 1964 to 1968.

24 Cinderella in Hebrew is *Lichluchit*, meaning dirty.

25 Operation Paul Revere was a 3rd Brigade, 25th Infantry Division operation that took place west of Pleiku, lasting from 10 May to 1 August 1966.

26 Israeli pioneer combatant youth organization.

27 The Hebrew spelling of this name can be interpreted with many English spellings depending on the vocalization, which is not given.

28 Nguyễn Phước Vĩnh Lộc (1923–2009) was a lieutenant general of the Army of the Republic of Vietnam during the Vietnam War.

29 Generals Matthew B. Ridgeway (1895–1993); Maxwell D. Taylor (1901–1987) (see chapter 1 note 57); Lyman L. Lemnitzer (1899–1988).

30 Colonel Marvin J. Berenzweig (1916–1996).

31 Summer festival of first fruits.

32 Likely Colonel Daniel H. "Dan" McGovern, who was the commander of the 3rd Brigade, 1st Infantry Division during 1966.

33 General Glen D. Walker (1916–2002).

34 See note 27.

35 Dayan must be referring to the Battle of the Bulge.

36 See Chapter 1 note 47.

37 Named after Vidkun Quisling (1887–1945). A traitor who collaborates with an enemy force occupying his country.

38 Dayan might have spelled this wrongly: could be Barclay.

39 By the time this diary was written, things had begun to change.

— 4 —

INTERMEZZO—'CIVILIAN ACTIVITIES.'

A Model Farm[1]
10 August 1966

This morning at 8.30 am I go out for my farm visit. A young man comes to fetch me: his name is Rufus Petri, son of a missionary family from 'Christian Assistance for Vietnam,'[2] instructing the mountain people in farming techniques on a demonstration farm. He grew up with his parents in Nigeria, and is very religious, admitting that he is a conscientious objector to the war. He is a bachelor, who wants to devote his life to developing countries, and is paid $30.00 a month, with board and lodging. He has studied at a college for agriculture in the United States and worked in an agricultural farm for several years. He is friendly and open. I find it interesting that both he and my cameramen from United Press, Tran Dai Minh, believe that, if free elections were held today (in South Vietnam), Ho Chi Minh would be elected. First—no nation wants war—it just wants to be left alone. The only people in favor of the war, and who support America, are war profiteers: the army chiefs, and businessmen who are making a fortune. But here—as in other developing countries—only such a group, or one like it, decides policy. I assume that it is the same way in India, Nigeria, or other similar countries where 90% of the population live very primitively[3] (who can expect mountain people, who walk around half-naked, to decide policy?). The 10% who represent 'public opinion'—members of parliament, journalists, and the intelligentsia—are the ones who decide foreign policy. The people's uniformity of opinion is striking. The cameraman also said that, if the Americans would leave (a completely hypothetical situation, to be sure)—the North would not need to conquer by force: local Communists would gain the majority, and the decision would be political, not military. In the beginning, those who supported neutrality, and wanted to reconcile the two groups, would have won the election. General Minh (at present in exile in Bangkok) wanted to reach such a compromise with the Viet Cong, and could have achieved it.[4] However, the Americans did not allow this to happen. But now, even that would not work.

All these are quotations from others. To properly gauge public opinion myself, I would have to know the language (and dialects), live here for a while, and—most important of all—gain the trust of the people to whom I was speaking. I cannot forget the stubbornness of the professor who refused point blank to say a single word, recommend a meeting with any of his friends who were not part of the government, or say anything about a visit to a village with them—all because of fear of subsequent punishment. At the end of the day, war is being waged here, and 'free speech' is shunted aside. This obviously also applies to the current elections, which—again according to the cameraman and missionary– are of no interest to the local population. Those standing for election are all government stooges and the voters could not care less who is elected. It is well known that Communists, and those in favor of neutrality, are forbidden from voting.

The start of our farm visit—several pigs for crossbreeding, an exemplary vegetable garden, and a sowing shop for young girls. The purpose of the farm is to teach pupils from nearby villages modern agricultural methods, artificial fertilizer, insect repellant, etc. The sewing shop is to teach girls how to use a sewing machine, for use in the village. The assumption is that the village will have at least one sewing machine. Several (male) pigs for each village (with hundreds of inhabitants). The farm's resources are very scanty: the few implements they have, come via American aid. There is no electricity, etc. There are five instructors, and six other workers.

I don't know whether their work has real results, but I believe their methods to be correct. They do not intend to initiate a revolution in the local economy, introduce modernization, mechanization, cooperatives, or any fundamental changes. Their goal is only to advance what already exists by one step. The raising of pigs—stock must be improved by breeding males with a better stock. Rice growth will improve with artificial fertilizers. (When I ask Petri if it would not be worthwhile to machine-dry the rice instead of spreading it out in the sun, etc.—he says no. It is forbidden to suggest something that will have to be bought with cash, to replace work by hands, which they have in plenty). He tells me that the pupils take vegetable shoots home and plant them. If he had ten to fifteen other similar farms in the area, he could really advance their agriculture. The meaning of 'agriculture' here is about 1 dunam[5] beside the house—for vegetables, bananas, cassava, etc.—and an additional 2–3 dunams for rice, per family, in areas outside the village. These areas are more concentrated, mostly found in river gorges, where they are arranged in a series of plots, to benefit from rain or wellsprings.

One of the interesting enterprises is to establish a village fishpond. Each village is located next to a well. Petri say that this is succeeding—they are raising carp.

At this stage, they want to improve the people's nutrition, not grow produce for sale. One day, if they want to do this as well, they will have to concentrate on several marketable crops: rice, coffee, tea (in this specific region), perhaps also cattle and pigs for meat. But this plan (which is objectively possible) requires thorough attention to marketing methods, transportation, etc. Transportation by road is difficult: travel is by convoy; there is constant fear of a Viet Cong ambush and no chance at all that the Vietnamese government—as it now stands—would participate in the project. The Americans seem to be satisfied with something between true philanthropy, and assistance only for the sake of appearances, to prop up the current agricultural structure. The young man hopes to receive additional aid from the United States—apparently for specialized pipes: perhaps they are connected to the mission. In any event, he is confident of obtaining 100 sewing machines, to achieve this specific goal (three girls are sitting in the 'classroom'; a fourth girl has taught them to stitch their national costume with a machine). I do not know whether the 'teacher' is a perfect seamstress, but she certainly could be elected beauty queen.[6]

The visit (that is the entire day) takes place in pouring, unceasing rain. I wrap myself in a poncho. Rufus rolls up his trousers, so as not to soak them in the mud (with a salary of only $30.00 per month, he must take care of his trousers) and becomes soaked to the skin (his shirt has no long sleeves)—but I see that he really doesn't mind.

When we return from the farm, two of Joe Hoal's assistants are there to greet us and tell us that plans have changed. Instead of visiting the villages, I am travelling (flying) to some coastal city or other, to obtain a briefing on activities of the American Welfare Administration [*sic*],[7] and lunch with the administrative heads. We travel to the airport; the plane is late because of the pouring rain. In the meanwhile, I overcome my hesitation to insult or not to insult and decide to insult. I say to my escort: 'I was briefed in Washington and will be briefed again in Saigon. I can receive tables and numbers in Tel Aviv as well—from the American Embassy. In Pleiku I want to observe facts on the ground. If you will not permit this—I won't do it, but I will be very disappointed. Tomorrow I am flying to Saigon, and I want to set today aside to tour the villages.'

My escorts aren't altogether surprised. The change in my original plan (which was a tour of the region) apparently came from on high. After taking counsel, they decide that one of them will accompany me on the visit, the other remain at the airport to report by telephone about my refusal to come for the briefing.

We go out to visit a refugee camp which is also a settlement area.

Refugee Camp and Villages

About 200 families hang around in tin structures. Each family has a room (without partitions), about 4 x 5 meters. In the room are two giant (wooden) beds, a coal-burner, and endless numbers of (dirty) children.

The refugees are supported by the government (I am no expert on the distribution of financial resources between what passes here for a 'government' and the different American foundations). They are supposed to be properly settled within 6 months. In this area, the counsellor maintained, the plan has worked.

We go out to the settlement area. On the way, the counsellor gives me his figures (in reality, they are far less good). The counsellor, Frank Favitz,[8] a full-time employee of Welfare Administration [USAID], has previously worked in other developing countries (he knows Haft, who worked in Iran for several years). He is no expert in agriculture but a general factotum. His only question for the refugees is whether they knew about the upcoming elections (11.9), and that they must vote for only one of many candidates.

The 'settlement' is something like a minimal apartment on a small plot, with communal toilets. This is what the government gives to each family:

1. For construction—twenty aluminum cans (the guide said twenty-five). The settler family must buy all the material for the walls (mats, dubious-looking cardboard boxes, frames of wooden and bamboo posts, etc.) themselves. According to what they say, the additional material costs 15,000 piasters extra. They receive the money as an advance on their salary from employers in the city.

2. Land. The guide says ¾ acre (about 3 dunams) for the house and 3 acres for the communal rice-growing area. We find 20–30 meters beside the house, no more. The guide apologizes, saying that land is exceedingly difficult to come by. Good land already has owners, and the government is not prepared to confiscate or buy it. The Saigon government would really prefer to leave the refugees to their own devices and not invest any money in them—let them work, and fend for themselves, they will manage.

3. The government gives them (work) implements. When we take a closer look, it becomes clear that the boxes of tools are from

American aid deliveries distributed any which way in the village. The refugees must buy blades and fix the handles themselves. They have a total of three work implements: a long-handled hoe for agriculture; an axe to chop wood; a machete, to cut the grass.

4. The government gives every family 3,500 piasters cash (settlers who are defectors from Viet Cong and North Vietnamese units receive 5,500 piasters—their policy of 'open-handedness').

5. Food for 6 months—rice and various kinds of supplements. The assumption is that, after 6 months, the settler should have his own harvest (from land, which is promised, but not given).

6. Communal toilets, and a school with five teachers for 300 pupils. Also, medical care: a doctor's visit approximately every week, and more regular treatment, whenever needed, by a male or female nurse. The Americans supply the medicines.

We visit a school where they put on a 'special performance' for us—they clap hands, cross themselves, etc.[9]—I peek into the schoolrooms, to see the children at their lessons.—A male or female teacher (most are female) writes something on the blackboard and all the children read it in chorus—with interest and enthusiasm. The children are all full of life, cheerful, frolicking in the mud, and laughing.

Mountain people settle the other part of the village, not Vietnamese. Their small plots (beside the house) are almost uncultivated. Sacks of artificial fertilizer distributed by the government lie one on top of the other, torn and rotting. Obviously, the fertilizer is wasted. The 'houses' – twice as many—contain two or more families. They tell me that this is the way they live in their natural environment. The room has no furniture. They sleep on a mat covered with some sort of blanket. They use forest wood to cook, with a few pots and pitchers of water, oil, etc.

Clearly, this is not a settlement. Anyone who wants to live here must work somewhere else or continue receiving a government stipend.

Apparently, there is work to be had in the area. The large American army provides work—building and services—and so the wheels of the economy turn each other, greased by the almighty dollar.

In sum, this is not a 'settlement,' but rather a kind of neighborhood[10] near the army base but far from the city.[11] Those that live here have a roof over their heads and a plot of land, on which the most diligent amongst them can grow some vegetables.

I speak with the refugees with the aid of 'my' Vietnamese cameraman. When ask why they had left their homes (villages), a chorus of answers replies: 'because of the Viet Cong.' I dare say that what they meant was—because of the war in the region. Simply put, they cannot live anymore in an area that is being bombed and full of troops. Their own villages are designated to be bombed if the Viet Cong enter them. I believe them when they say that they don't want to return—even when peace comes. I am not sure what they imagine when they say 'peace'—they have already been living under war conditions for 20 years and, as far as they are concerned, 'peace' surely means a kind of temporary relaxation in the war. Furthermore, here they are sure that they will receive food from the government (it is likely that this distribution will cease, but in the meanwhile—it comes). There is work with payment in cash, a school, medical care, and a bit of land. In total, I am doubtful whether their conditions were better, where they lived before. The main thing is—rice, which the government hands out free of charge.

I give a 'test' to a little boy rubbing himself between my legs—I ask him to write his name down in my notebook—he does so enthusiastically and effortlessly. After that, I write 2+ 2=, and he immediately writes down 4! Bravo! (Rufus the young missionary tells me that Vietnamese are excellent with their hands: they can dismantle a machine perfectly and then put it back together again like new).

Because of the change in my plans, I remain without an escort from 1.00 pm. I suggest that we return and look for Rufus. We are lucky: when we return to the farm, Rufus is nodding off to sleep over his Vietnamese grammar book. He agrees to obtain a car and accompany me on a village tour. We have lunch in the city (Pleiku) and then go out to the villages (rain, rain, nothing but rain). All the way there, half-naked women, babies or baskets tied to their backs, walk to and from the rice paddies, weeding, adjusting the water, etc.

The villages are built on straight road after straight road; the houses are built on top of stilts and are entered by way of a 'wooden ladder'—a tree-trunk with notches. One must be barefoot to be able to lean against the rungs; my army boots are unsuitable for this.

The houses are built high up, to isolate them from water and ground dampness. And they really are completely dry. A group of villages is surrounded by a canal, into which sharpened bamboo branches are embedded—a kind of wire fence to defend against Viet Cong attacks.

In one of the villages, two local 'soldiers' stand at the village entrance, armed with rifles. The village economy is based on 1 or 2 dunams beside the houses (fruit and vegetables, some cows and pigs, a few chickens), and rice paddies in the nearby stream.

Outside the village, or adjacent to the house—are the rice depots: small wooden buildings with tin roofs, also on stilts. The rice is kept there all year and can even be dried there.

After Rufus has 'examined' two sick children (he hasn't a clue about medicine), and promises to send them a doctor as soon as he can, we return to camp. On the way, we manage to sink into the mud, and must push the car forward and backward, until we extricate it.

To sum up, it seems as though the Americans are doing the right thing. With very little money, they will succeed—if they persevere—in raising the villagers' standard of living, improving their nutrition, and providing them with primary services such as basic schooling and health care. (In the refugee village, the teacher complains that they don't have books or writing paper. The instructor explains to me that all this is available in Saigon but is difficult to transport to the villages because the roads are dangerous. What gets there is mainly by air).

I see no signs of real development—in other words, fundamental change. This would be as difficult to achieve as parting the Red Sea, and require proper marketing organization, preparation of large additional areas for cultivation, mechanization and modernization. This cannot be done without a fundamental revolution in their way of doing things. In today's Vietnam, I doubt that they have either subject or object: who will start the revolution, and who will be prepared to adapt to it (the farmers).

In two villages, I see the product of crossbreeding local pigs with high-quality pigs from the United States. There is something in this, in principle. But I don't see (apart from a few cross-bred pigs) anything in the villages resembling results obtained on the model farm. I didn't visit the rice paddies—perhaps there is some change there (because of artificial fertilizer).

'My' Vietnamese cameraman tells me 'in secret' that the American Welfare Administration [USAID] are all spies ... They call this 'discovering America.'

It must be said in favor of the American instructors, that they are young, enthusiastic, serious, and do not get worn out by despair because of the difficulties they encounter. They are also not 'sophisticated' enough to compare the faded chauvinism of the 'forester' Robert Daver from Oregon[12] with the personal, idealistic approach of Thomas Sturdevant from the international student volunteer service.[13]

I am very interested in the declared orientation of the American developmental plan, as expressed in documents received from Joe Hoal.[14]

The refugee settlements are nothing like serious agricultural settlements. Perhaps it is an error, and loss of a historical opportunity, for the Americans to establish a model settlement here, to study the problems and really open a new page in Vietnamese farming—It will be a much more difficult task when they must improve existing farming methods. They are missing a golden opportunity to start from scratch.

In the meanwhile, I have been informed that the company, with which I was supposed to go when they landed (I was later transferred to the aid company—Barnes's [*sic*] second company), has almost all been killed (they came upon Viet Cong bunkers next to the river). Interesting: what would have happened to me if I had been with them?

Back to the War

During dinner, Colonel Lee tells me that last night there was a large battle involving a Korean unit 5 kilometers from the Cambodian border). The unit—company—was well dug in. The North tried to defeat them all night—and failed. Two hundred bodies were left near the unit's barbed wire fence. The unit was helped by artillery fire, tanks and floodlights. Wave after wave of North Vietnamese returned and did not stop until they had been all killed. They were shaved, had their hair cut and were dressed in clean clothes—in other words a unit which was rested and ready to attack.

After what he said, I tested some of my thoughts on myself and my table companions.

1. Why did the American company that was hit so hard the day before yesterday advance all the way to the river? A company is not the same as a patrol looking for contact. If there were no Viet Cong in the vicinity—why go?—They should have dug in first and ensure artillery back-up, giving the enemy an opportunity to attack—as the Viet Cong have done in most cases, when the Americans have landed near them. In all cases, one must distinguish between 'search' and 'kill.' Search is not carried out with a company, but in small mobile units, who remain stationary in the area for days at a time, to observe enemy positions. A larger unit (company)—landing by helicopter—is easy to see and hear. If this company, instead of digging itself in—goes out on patrol, it is liable to encounter an ambush or fortified position and suffer serious losses. This would have happened to the Korean platoon, if—instead of digging in—it had gone out on mobile patrol.

2. The eternal question is—why they don't let the Viet Cong concentrate in large units (divisions) and prepare their 'monsoon attack'—using the opportunity to destroy them. The current method of searching and trying to destroy the Viet Cong wherever they are—leads to war against small, divided units, who often succeed in splitting up even more, scattering throughout the area.

3. If things carry on as they are now—with the Viet Cong scattered everywhere, and Americans pursuing them wherever they are—the Americans will be compelled to bring in more and more troops, to militarize the entire country. From the Vietnamese population's point of view, the army would become an army of conquest. The Americans would be prevented from forming a line and fighting a decisive battle, and instead of defeating the Viet Cong, all they would do is split them up into even smaller units.

I am surprised by my support. The commander who preceded Norton[15] wants to fight this way. But Washington, and certainly Westmoreland,[16] is not prepared to do this. If a division commander—for example Larsen—would let the Viet Cong operate in concentrated groups, Westmoreland would immediately call him up and ask why he was not pursuing the Viet Cong forces discovered in the area. If I understand correctly, there are two reasons for this American tactic:

1. Military-emotional: Americans are aggressive, full of energy. Almost instinctually, they pursue the enemy, strike and fight in every place the enemy may be found. When the enemy cannot be found, they look both for him—and for battle.

2. Double politics—from both Vietnam and Washington. In a Vietnamese context, they cannot allow themselves to be seen allowing the Viet Cong to concentrate in, and (temporarily) rule,

> areas, with the assumption that they will be hit and destroyed later. Every time the Viet Cong makes any temporary, local progress, there is immediate negative political interpretation. The American public—the press, Congress, etc.—would react badly to such a tactic. Television broadcasts, and continuous press reportage on every battle, do not permit patience or faith in army leadership to either order withdrawal, or stay in place. The press hates inactivity.

I have no doubt that the Americans would win, even with this method of fighting. They can allow themselves to pursue two Viet Cong with ten helicopters for days at a time—and eventually catch them. But they themselves are contributing to the Viet Cong splitting into small units, thereby prolonging the war and obligating themselves to take up artillery positions throughout the country. This necessitates enlargement of the army and creates a negative impression on the population. War everywhere, and the American Army (and all its negative consequences) everywhere.

I ask whether I can pose an irritating question—of course, they say go ahead. Do they really believe that this war will proceed in brilliant, textbook fashion, deceiving the enemy and conquering it quickly; or does this simply fit in with what they prefer to think? They all admit that none of them has any tactical experience, but, because of their force-power ratio, they don't need any.

Although I understand their starting point, I am not convinced that they are correct (I am also not convinced that the opposite is true—that they are making a mistake). It also depends on the terrain. The Viet Cong have no rapid means of transportation, and if it were possible to draw them out, letting them concentrate and attack targets outside the jungle—they could be hit afterwards in open terrain, because they would not be able to retreat rapidly. As far as the current situation is concerned, I have no doubt that the Americans will continue to slaughter the Viet Cong. It is possible that Ho Chi Minh is too far away (I am told that the Viet Cong only have radio communication at battalion level, and that the Americans receive and decode their transmissions, they don't understand the situation properly and continue ordering units to attack). But local Viet Cong commanders also do not understand that, when they see a Korean company that appears isolated, they can call in the aid of American artillery, air force, and armor, with all their up-to-date communication and traffic facilities. Any attempt to subdue such a unit would, in my opinion, fail.

I am not surprised to see, in today's newspaper, Nixon's[17] announcements on the need for more troops in Vietnam (he had first been briefed by the army commander), also a notice from the Pentagon that they would not hesitate to send out more soldiers, if required. These two reports reflect the perception of those in responsible military positions, and the fact that Vietnam has become a political issue in the forthcoming elections. Unlimited aid for 'our young men fighting in Vietnam' has become a convenient election slogan.

I receive a letter of thanks as well as words of praise from Norton for my book, and a request to return here after Saigon. He pleads with me to return: our conversations at table have been very instructive and he wants to continue them. I take a different, non-American, approach to many issues, foreign policy, etc. He suggests that I come here to have a rest; he sees that I get no rest from early in the morning till late at night (writing).

At least for now, there is another possible answer to the fundamental question—on American tactics (and perhaps strategy) of the war in Vietnam. I have hinted at it already: allow the Viet Cong to concentrate and attack, and by this means create formal battle lines which prevent them from scattering—and defeat them.

12 August 1966
Saigon

This morning, I visit Cam Ranh Bay—one of the world's best natural bays (deep water, large, enclosed). The Americans are making this into a central logistical base. Two Air Force wings are currently located here:

there are 150 F4Cs ('Phantom')—a plane that pilots 'love' because they see it as a MIG-destroyer—and about 6,000 men. Together with other maintenance units—a total of 12,000.

U Thant[18] and others have mentioned this 20,000-man base as an example of the Americans' desire to really settle here. I don't know how many hundreds of millions of dollars have already been invested in this base. It is doubtful if anyone knows when it will be completed (usually places like this are built and built, with no end in sight). In any event, according to what was explained to me today, the plan is for a permanent, large-scale construction. I ask whether the United States has any ownership agreement for this area. I receive no clear answer, but it is very likely that they have no such 'title deed,' and do not worry that they will be thrown out, because of mere lack of an ownership certificate. A leader like de Gaulle and country like France are necessary to tell the Americans—one of these days, you will have to leave here.

Huge amounts of ammunition, fuel, spare parts, other equipment, and food have already been concentrated in the camp area. There are many ships in the harbor, and amphibious devices are being unloaded from ships far from shore. The equipment comes by sea and goes out by air. There are currently no main roads in Vietnam that are freely open to traffic, without fear of interruption by the Viet Cong.

The Air Force commander Colonel Levi Chase[19] (A Jew?): has already made 500 sorties. While we were walking round the airport, a phantom jet returned because of radio trouble. A Jewish young man, Klafnitzki, hears my name and asked whether we need pilots in Israel: he is prepared to come. (I ask him: can you bring your plane along?).

While I am being given explanations about the planes, the pilot tells me that their electronic disruption of enemy ground (SAM) missiles works perfectly.—If used correctly. If not—it is not because of incorrect methodology, but incorrect operation.[20]

They are proud to show me the radio used to call for support, and_bombs with an umbrella that opens in their fins (which makes them drop down to earth in a straight line, not endangering the plane when they explode), and other equipment.

It is probable that every farmer, both here and in the rest of Vietnam, could have a bulldozer to work his land, from all the surplus equipment remaining here. (It is surely so with all American wars). The main thing—is to operate rapidly with plenty of whatever is need. Equipment—unlimited.

The only camp in the area fenced in with double and treble barbed wire—is a collection of service huts, that work in isolation There is also a military police building here, for extra security,

There is a Vietnamese village at the edge of the camp—as usual, full of women of all ages. There is no village fence, and the constant flow of soldiers there testifies to its popularity.

Amongst its other features Cam Ranh also has high mountains in the middle of the enclave, allowing radar cover for the whole of South Vietnam.

Dinner with the one doing all the explaining. John Stuart, with two other men: J. Levine and one other colonel. They ask for my opinion—especially if I have any criticism. I do not hold back and express my critical opinions. A lively discussion develops—at times even more than that. The main points are:

1. Refugee settlement

    a. The Americans are missing a historic opportunity to build properly planned, modern settlements. This is difficult to achieve, with agriculture that already exists. In Iran after the earthquake[21]—villages were destroyed and could be built from the beginning in a modern, planned form. Something like that could be done here as well. They are missing this opportunity—to start from scratch.

    b. The problem with the refugees is partially their own fault. These farmers are running away from the war, not from the

Viet Cong (with whom they could come to a compromise)—they have been made into refugees. This is an American war against the spread of communism. Vietnam, on its own, would not have fought this war. I don't know what Vietnam prefers—Ho Chi Minh, or the war. Today's newspaper states that there are about a million refugees, and each month 20,000 new ones are added. This increase comes from the way this war is being waged—throughout the entire country. Wherever there is bombing and artillery fire—the population have fled, flee, and will flee. The Americans are responsible.

c. The Americans cannot be required to settle all refugees: the country is at war. But in places where they do settle them (a total of several dozen villages), the villages they erect must be exemplary—instead, they are establishing slums. In Venezuela, settlement of one farming family costs $11,000.00—here, the Americans invest less than $50.00 per family (with additional food for several months). Under these conditions, two things must be done: decent housing (minimal, but not slums) for those working in their camps, services, etc.—this comprises only a few of the refugees. Proper housing as well for those who are settling on the land. A simple two room dwelling, 1 dunam beside the house and 4–8 in the rice-growing area. A few pigs, cows and fruit trees, a village fishpond. By doing this, they would raise the farmers' living standard a dozen times over and give an example for others to follow. There are no marketing problems here—rice, meat, etc. are imported from the outside, and the population's standard of nutrition is very low. There is no problem with land—proper preparation with heavy mechanical equipment can make the regions ready for the planting of rice, vegetables, tea, coffee, etc. But what do the Americans present to the world? The Welfare Administration's [USAID's] address, and miserable tin shacks. It would only cost the price of a few shells, to do everything I have enumerated. The Americans are making a fatal mistake every way they can, wasting a historical opportunity. It would be simple for them to erect showcase village and neighborhoods, thereby conquering the hearts and minds of the inhabitants. This is both an opportunity and a duty.

2. Publicity and propaganda. They treat the whole world like children, scattering slogans, and expecting everyone to accept them.

    a. This is an American war, in America's interests (United States interests are global). Their maxim that they are only here 'to come to the aid of the Vietnamese, and when asked to leave—they will leave'—is incorrect. Which Vietnamese are being asked? How are they asking? By the upcoming 'elections?' Is this the free will of the people? If the people here really are tired of war and want it stopped—are the Americans prepared to stop it?

    b. They have a series of slogans 'for every occasion.' For the army—'search and destroy'; for those involved in aid work—'to win hearts and minds.'

    c. President Johnson says to the world: 'Our goal is not acquisition of land, but to come to the aid of our allies'—he says to the American people that America 'acts as a floodgate against communism and its international emissaries.' When convenient, they refer the issue (Vietnam) to the UN. When inconvenient—they 'twitter' about the UN. 'Chinese expansionist aims'—this phrase requires closer scrutiny. They organize elections—and afterwards assert that those elected express the will of the people—but elections are not free or fair. In the present (wartime) conditions, if free elections cannot be held—it is better not to have them at all.

3. If the war does not end within a year, the very fact that the Viet Cong have held on during such conditions will be a kind of victory. The Americans have 500,000 soldiers, an air force, artillery, communications, convoys, armor and unlimited money. And even with the million-man Vietnamese Army, they cannot defeat this wretched army of barefoot soldiers, without any real equipment. China and Russia are both not getting involved (with manpower). And still, the American Army cannot find a way to defeat Hanoi and the Viet Cong and end the war—especially when their method of war demands that it be waged throughout the entire country.

4. Large patrols (in companies) do not serve their purpose. A small number of men, remaining in the area for a few days (if possible—in position) would find the enemy more effectively. They must break off initial communications with the helicopter (by distance and time and, when they land, each one must know exactly where they are). They must clearly distinguish between 'search' and 'destroy.' After the small, immobile patrol finds the Viet Cong—they should inform the unit, which should arrive in strength as soon as possible (perhaps the frequent replacements of American soldiers prevent them from acquiring the expertise necessary for such patrols).

I don't think that Ho Chi Minh knows the exact situation on the front. Perhaps he remembers how things were done during his war against the French—who also had an air force, artillery, etc.—and assumes that he can do to the Americans what he did to the French (like all generals—he remembers the previous war). If this is not so, I have no explanation for an example such as the Viet Cong's suicidal attacks against the Korean company. The Viet Cong authority deciding war tactics are too far away from the front (these are my own assumptions, to explain the Viet Cong's behavior).

5. For some reason, the Viet Cong call this a 'guerilla war.' But it is not so. At this stage, the North Vietnamese forces are conducting a 'regular war,' to drive out the Americans; however, the way the Americans are conducting the war is causing the Viet Cong to split into small units.

Tomorrow at 5.30 am I will go out into the 'field.' If everything goes as planned—and there is no special reason to lengthen or shorten my time there—I will return on Thursday morning 18 August, to meet with Ambassador, Henry Cabot Lodge,[22] at 4.00 pm.

I take a hot shower and operate on ('burst') all kinds of suppurating contusions, 'gifts' from the jungle—mainly ant bites and thorn scratches.—To stop infection, I spread shaving cologne all over and feel better.

In the meanwhile, the Viet Cong have sabotaged the electricity supply, and there is insufficient electricity in the town. The air conditioning has stopped functioning, and we only have candlelight in the room. Of course it is stiflingly hot (to increase the irony, a picture of snowy Switzerland hangs in my room: the right picture in the right place!).

The jungle is a ferocious beast! It can bury all the Americans and Vietnamese in it, enveloping them with bamboo from below, and monsoons from above; and no one would know that anyone had entered it.

Endnotes

1 In this chapter, Dayan is referring to the "Strategic Hamlet Program" established by the government of South Vietnam in 1962 with advice and financing from the United States. He does not use the phrase "Strategic Hamlet," instead calling them "refugee camps" and "settlements."

2 There were quite a few Protestant and Roman Catholic missionary and humanitarian organizations working in South Vietnam. The title of this organization is unclear as given by Dayan.

3 Translator's note: An extremely racist and patronizing thing to say.

4 Dương Văn Minh (see Chapter 2 note 14).

5 A dunam, also known as a donum or dunum and as the old, Turkish, or Ottoman stremma, represented the amount of land that could be ploughed by a team of oxen in a day. In Israel it is equal to approximately 900 meters.

6 Again, Dayan's appreciation of beautiful women.

7 Here, Dayan certainly is referring to the U.S. Agency for International Development (USAID).

8 The name is uncertain, it might also be Fawitz.

9 Approximately 7% of Vietnamese are Roman Catholics.

10 Hebrew *shechuna,*

11 Dayan does not refer directly to Strategic Hamlets as such.

12 Robert Daver, Oregon forestry expert.

13 Thomas Sturdevant was known for his work in international volunteer service, particularly in the realm of forestry and environmental conservation. His contributions were notable in helping communities around the world through volunteer work that focused on sustainable land management and forest conservation. Sturdevant's efforts typically involved working with local communities to implement forestry practices that promote environmental sustainability and support local economies. His volunteer work might include projects such as reforestation, forest management training, and the development of community-based conservation programs.

14 See Chapter 3 note 27.

15 Chapter 3 note 18.

16 See Chapter 1 note 50.

17 Nixon was out of politics at the time.

18 See Chapter 1 note 44.

19 Major General Levi R. Chase, USAF (1917–1994).

20 No technology is foolproof, and this may have been enthusiasm by the pilot. There were different versions of the Navy and Air Force Phantoms and countermeasure systems were varied between the services and the models of F4s throughout the war. 1966 was still early for the American experience and the technology was not in the same condition as it was later in the war.

21 The 1962 Buin Zahra earthquake, 7.1 on the Richter scale, resulted in 12,225 fatalities.

22 See Chapter 2 note 2.

— 5 —

BACK WITH THE SOLDIERS

'The Green Berets' 13 August 1966 Duc Kho Camp[1]

A 'Special Forces' and 'Civilian Irregular Defense Group' (CIDG)—camp.[2]

(They tell me that these are civilians, irregular—very defensive—nothing else—a group—I can get nothing else out of them).

I arrive with Colonel Farmley, who is responsible for this camp. His commander is Colonel Kelly, under the supervision of Major General Larsen.[3]

I overnight here. I have been here once before, but only for a few minutes, when I was with the 1st 'Cavalry' Division. The mud is waist high (I have seen mud; during the first years of Nahalal,[4] we waded in mud up to the knees. But never in my life have I see mud like this—'thanks' mainly to the tanks that churn up the ground already wet from unceasing monsoon rains).

The 16th American Armored are active in this camp, under the command of Captain Cunningham; there are also 700 conscripted Vietnamese mountain people.[5] They do some patrolling in the vicinity, protect the nearby villages and—principally—form a firm base next to the Cambodian border (The book *The Green Berets*[6] was written about this base). They have good fortifications, bunkers, subterranean rooms, some artillery and mortars. But their main importance is communications equipment, with which they can rapidly call up artillery and Air Force assistance.

There is a nearby (non-asphalt) airfield, a provisions and significant artillery unit and an armored (tank) division.[7] They supply intelligence, instructions, and other services to units operating in the vicinity. (At the moment there is a Korean unit nearby). The Americans are subordinate to the Vietnamese Army in order of obtaining support; they are, after all, only 'advisors.'

Altogether, there are more than 30,000 of these 'special forces' (mostly local Vietnamese and mountain people), dispersed all over the country—mainly on the Cambodian border. The value and place of this unit must be appreciated within the framework of the entire American campaign. They provide local services, regional expertise, and secondary guard duty (road convoys, etc.). By these means, they preserve the manpower of significant battle units. On the other hand, their patrols are without question amongst the most dangerous, also because of their proximity to Cambodia. (The camp is in the middle of the 'Ho Chi Minh Trail,' where it crosses from Cambodia into Vietnam. Of course, 'trail' stands for an entire area, the central route by which infiltration occurs, not a real, narrow road); for this reason, as well, they work in small groups.

The 'special forces' camps have already been overcome several times by the Viet Cong (more accurately the North Vietnamese Army), mainly at times when they were not under American command. It is impossible to depend too much on the local military because their defenses can be breached. The captain mocks their bamboo obstacles, (showing how he makes them fly away with one good kick of his boot, opening a road for himself without difficulty).

In any event, American soldiers sent to these units are special volunteers. Their adventurous instinct and strength of spirit have acquired a good name ("The Green Berets') for them in the fighting unit.

Their commander, Colonel Kelly[8]—is a 'character.' A New York Irishman, who knows all about Judaism because of his Jewish neighbors (he proudly tells me that his uncle was killed as a rebel by the British during the Irish Rebellion). When I tell him that the British had created a special place for me in their prison—he says: 'who knows, we could have met there: Irish and Israeli rebels together.' Colonel Farmley also has an 'association' with us—his father was in the same class as David Marcus,[9] and is buried in West Point.

I reach Colonel Kelly via General Larsen. Kelly explains his unit's structure to me, also his assessment of how things are developing in Vietnam (he is an optimist—it will take another 15 years to finish this war) and hands me over to Colonel (actually Lieutenant Colonel) Farmley, to 'tour' the units with me.

During our helicopter trip, Farmley explains the deficiencies of the American submachine gun in jungle conditions (usually this gun works well, and everyone praises it):

1. The magazine usually works well, but if it bumps against a tree it may fall without anyone feeling it.
2. Its gun-split side of the barrel (barrel mouth) gets entangled with climbing jungle tendrils.
3. The switch that regulates firing speed can be set at locked, single fire, automatic fire. In the jungle it should be set on automatic, especially after having been 'locked,' when fire is really needed.

Captain (Cunningham), Colonel (Farmley), and I have dinner together. After beer, a lively, open conversation develops. The scarlet thread running through everything that the two American officers say, centers on two issues: The first—self-confidence, pride, a feeling of superiority (over the rest of the idle world), faith in their success, their weapons and the righteousness of their cause, The second—lack of clarity on how victory can be achieved, and the war ended. This lack of clarity assumes many—sometimes acerbic—forms and is expressed in criticism of (their) political leadership who do not allow them to complete the job, (conquer Hanoi, use nuclear weapons, enter Cambodia and Laos, increase the number of soldiers here, etc.). This also influences individual campaign issues, for example opening the network of roads for routine use, or at least in convoy. Absence of open roads is a burden on the economy (it really brings it to a halt) and is symptomatic of the absence of peace and lack of progress in 'pacification.' Just today, two colonels expressed great hesitation about whether anything can really be achieved soon, in this area. Many soldiers are needed to protect the roads from Viet Cong ambush, clear away mines, etc. (mines are being laid even in the roads that are currently in use). Furthermore, open roads will also increase the opportunity for guerilla attacks: the main users of these roads will be the government and the army, and the Viet Cong will not hesitate to strike at them: thousands of kilometers of convenient attack objectives will be added.

My escorts are pessimistic about 'pacification' of the mountain tribes. When they leave the region—the Viet Cong immediately return. The war that goes on here without a stop (on the central heights of Pleiku) has turned the place into an unpopulated no-mans-land, and they see no chance that this will change soon.

The famous Korean company that killed about 200 North Vietnamese soldiers (who had attacked them)—compared to only eight of their own killed—is operating in the vicinity (5 kilometers from here). They had been dug in in place for 6 days, and thorough entrenchment saved them. The prevailing opinion is that, had this been a company of Americans—who don't dig themselves in as well (head camouflage, etc.)—they would have been overcome in this powerful attack.

From the air, we see entire areas of bombing craters that look like land in preparation for tree planting. Even at night, the shock of bombs and cannons does not stop even for an instant (21,000 shells were fired during the night hours, in defense of the Koreans).

I start the morning at 5.00 am (Sabbath!).

I get into the plane at 6.00 am, but at 6.30 am, the colonel tells me that the pilot (a Black man) has politely informed him that we should board a reserve aircraft. We finally take off at 7.30 am. The previous evening, apart from writing (and before that a visit to Cam Ranh Bay, dinner with Stuart, etc.), I do some serious laundry in preparation for this morning's trip to the field, where I will remain for several days. To avoid eventually having to pay the Hotel Embassy's obscenely inflated prices, I turn my room into a laundry.

In the plane—even before work starts—I am therefore already tired and wrung-out. The sergeant in the cockpit saves the situation (I sit there as a guest of honor): he prepares hot cocoa from American battle rations. This is really a 'big deal'—all that I still need is whipped cream!

At 10.00 am, I meet General Larsen in his study. I have heard a great deal about him already, especially his aggressiveness and the name that he had acquired for his units: 'the fighters.'

His external appearance surprises me: lean, not tall, the face of an intellectual (his form reminds me of Wingate, and his speech of Montgomery):[10] clear, measured, short, unambiguous sentences.

He converses willingly and clearly enjoys the opportunity given him to develop his theories. The conversation starts with my plans. I tell him that I want to hear an explanation for his 'philosophy' of war here, and after that to visit his units, talk to their commanders, and observe how they operate in the field, join in patrols, etc. He says Westmoreland[11] has told him that he is responsible for my security during this visit. However, when I tell him that I have already gone out with both Marine and 'cavalry' units—he agrees that I could do this here as well. After that, we start to talk 'philosophy.'[12]

As far as unit operation is concerned:

1. Patrols: For small posts that have the capacity to summon large units (or operate themselves from ambush), on condition that they are sure of obtaining artillery back-up (each operational area is covered by artillery, including '175's with a 30 kilometer range (?). Strategic bombing—with B-52 aircraft (that bomb by radar and are weather-independent).

2. Patrols must not get too near the Cambodian border, and must leave place for future units to arrive, surround, and block off the road through which the Northern Army retreats. He complains that General Walker does not listen to him and gets too near the border.

3. He does not think it advisable to penetrate Cambodia—however, they may enter Laos, because this is what Laos wishes.

4. Neighboring countries: Indonesia, Thailand, Laos, Cambodia—have been psychologically strengthened, and taken courage from, the Americans fighting against the Army of North Vietnam, to overcome the threat of communism in their midst. The Western world neither understands nor appreciates what the Americans are doing, but history will vindicate and thank them.

5. The United States erred at the end of the Second World War by helping France return to Indochina and the Dutch to Indonesia. By doing so, it became identified with other imperialists. America should not have done this.

6. De Gaulle is a great French leader, but he will bring his people to the edge of the abyss. He is doing all he can to subvert, insult and hurt American relations. Here too, he incites against America and wants them to be defeated because he (France) was defeated here.[13]

7. Dictators who have recently popped up have no idea what the word 'economy' means nor of how to lead a nation—Ne Win (Burma), Sukarno (Indonesia), and Nasser[14] have dragged their countries into the mud, causing economic chaos. They are not capable of leading their nations.

8. The situation here has improved during the past year. The North Vietnamese Army is less sure of itself and not rushing to attack. Their warehouses are short of food and ammunition. Some of their soldiers have been captured barefoot (captives, wounded, etc.).

9. The bombings in the North are having the desired effect. He remembers the Germans in Berlin and the Japanese after their defeat: a terrified people. Continuous bombing is very depressing. The evacuation of Hanoi must surely have been a hard blow for North Vietnam, and cessation of traffic has destroyed its industry, etc.

10. Their soldiers are also having a tough time. Malaria is common, they are malnourished, and must travel on foot for months, packs on their back. The Army of the North is in a depressed state.

11. The Northern Army has real volunteers. Several days ago, a young, handsome, honest-looking Vietnamese corporal was brought to him. He was shaved, his hair was cut, and his clothes were clean. Larsen asked him how he had become a corporal at so young an age. He replied that he had enlisted 4 years ago, at the age of 16. Why? 'I volunteered to free my country from foreigners.' 'Did you desert (or give yourself up), or did you fall into captivity against your will?' 'I was taken captive. If I could, I would return to fight.' Larsen added that he would like to have such a corporal fighting for him.

12. Ho Chi Minh and his gang err, when they think that war with the United States is the same as that against the French. But by now the difference has become clear, and they must surely understand the situation. Therefore, it cannot be excluded that one of these days, Ho Chi Minh will decide to stop sending his army to the South, whether he admits it or not. Then pressure to stop the bombing of North Vietnam will build in the United States, because the North has stopped fighting. Larsen himself is of the opinion that the bombings must continue, to force Ho Chi Minh to the negotiating table, and sign an agreement not to cross the seventeenth parallel; this is the object of the war, and only this will be its solution.

13. He believes that the United States must not, under any circumstances, cross the Cambodian border. The political damage is not worth the military benefit. In the same way, it is not permitted to conquer North Vietnam, because then China will be forced to come to Ho Chi Minh's aid by entering the war, which will then become nuclear—something the United States must avoid at all costs.

14. The expected so-called North Vietnamese 'monsoon attack'—is nothing but an anodyne to be fed to the public. He does not believe that the army of the North will ever concentrate its forces in division strength and attack. His formula: during the monsoon, the North will concentrate their army and increase its force ratio, but not attack in strength. He agrees that they must (it may even be possible to) find ways to lure the North into open country, allow them to attack in force—and then destroy them. (He smiles: perhaps he doesn't want to speak clearly).

I enjoy the conversation, especially his cold analysis, which testifies to the fact that the only solution to the Vietnam question is the signing of an agreement that the North will recognize the seventeenth parallel and promise not to cross it. I do not know whether such a thing is possible. I have no way into the North Vietnamese psyche.

It will be interesting to see whether, after the severing of rail transportation, the Soviet Union will feel obliged to send supplies to Vietnam by sea, and whether this will lead to a confrontation with the United States. I assume not. Neither side is eager for such a confrontation, nor will the Soviet Union go to war with the United States—over North Vietnam.

The refugees in this area have crossed into Pleiku, with no intention of returning. Their number has increased—and the government is forced to support them (rice is brought in by air from the United States). The refugees like this—they move into an urban environment, near the seat of government—so why should they return to their villages? If this situation continues, the number of refugees streaming into the source of government support will increase every day.

14 August 1966
With the 101st Brigade in Tuy Hoa[15]

The commander of this brigade is General Pearson[16]

I arrive here from Duc Kho [*sic*].[17] We stop on the way at a 'special forces' camp at Cung Son.

I meet General Larsen here as well and arrange that I come to his headquarters on Wednesday evening for dinner. (If all goes as planned, I will return to Saigon on Thursday morning, for my meeting with Ambassador Cabot Lodge.)[18]

I am shown to my 'room,' a tent with a bed in it. But the main thing is—electricity, which I am very happy to see. General Pearson invites me for a conversation 'to boast about his activities' (he smiles when he says that) at 5.00 pm. Meanwhile, from what I hear with half an ear they are starting a large operation tomorrow.

Pouring rain this morning (and the whole of last night). Last night we chatted (with the camp commander Captain Cunningham) until midnight, about everything under the sun. The real importance of my visit to the camp of the 'Green Berets' comes not so much from what I see, but from my discussions with commanders—mainly with the sergeant (perhaps he is an officer) in charge of intelligence whose job it is to interrogate the prisoners.

Our departure is delayed until noon by pouring rain, and after that another rainstorm diminishes visibility to absolutely zero.

This morning, a patrol of two companies leaves the camp (200 men; two or three Americans with each company for communications, etc.). The patrol will remain in the area for 4 days. One company is to secure the landing area, the other to patrol the vicinity. If there is an encounter with the Viet Cong—they will summon the 'cavalry' to reinforce the landing area.

The departing company looks a little strange. All ages (from 15 to at least 60), loaded with large knapsacks (in addition to military rations, some of the locals also take a bag of rice); the Americans stand out amongst them like Gullivers amongst tiny Lilliputians.[19]

This 'special forces' camp is an 'Institution' with its own atmosphere. The gates are locked in the evening (double locked, with two mined fences). The shutters are closed, and the men go into camp for the night. The army rests. Outside, torrential monsoon rain pours down without stopping. The commander Captain Cunningham—after a prolonged discussion with me about everything thinkable—strums American folk songs on the guitar (to himself, everyone else has gone to bed). From the other side of camp, the Vietnamese unit hut, I hear a melancholy trilling. It is as if the rhythmic salvoes of the gunners (one every few minutes) accompany these lyrical-nostalgic outpourings. A paradox—guns contribute to the atmosphere of peace in the camp, a peace based on security, allowing the soldier to cover himself with his blanket and sleep, safe in the knowledge that someone is protecting and defending him. The night rest of an American camp in the Viet Cong jungle.

I manage to absorb the shots from the 105 mm cannons and continue sleeping, but the 155 mm barrages shake the walls, and everything hanging on them clatters. I have no alternative—and take three sleeping pills.

Although we are supposed to leave at 10.00 am, we only leave after lunch. The first station that we go to is another 'special forces' camp. There are supposed to be six companies (600 men), thirteen Americans amongst them; but now the complement is below par. A few days ago (on 10.8) its forward position was taken (by the army of the North or the Viet Cong). About thirty men were killed, and a few disappeared.

In fact, according to the words of the commander (captain), the matter is more complicated. The government (both the local government and the Americans) is unpopular because they have not supplied the population with needed foodstuffs (there are about 30,000 people in the valley). Two years ago, this region produced about 600 tons of rice, 200 for domestic consumption,—and 400 for export. During the past year, the Viet Cong has taxed (in other words collected) 80% of the harvest, leaving the people only about 60 tons. The government has promised to send sufficient food (rice) but not done so. (This year, the farmers did not plant most of the area, for fear of not being able to collect the harvest). Before the Americans came, when the area was under Viet Cong rule, they reached a kind of 'modus vivendi.' But when serious fighting broke out here between the government and the Americans, and the Viet Cong, 'the balance was broken,' and now the Viet Cong takes what it can—and it takes a great deal.

With this background (or that of information on an approaching Viet Cong attack), about 200 Special Forces men have deserted. They don't take their weapons with them—they simply run away and do not return (this is not considered a serious transgression). The camp is left almost defenseless—there is only one company in the area on patrol. The Viet Cong exploit—or plan—this situation of weakness and take the camp in 20 minutes (their presence is not felt until they explode a Bangalore torpedo under the forward position fence). Most of the soldiers (including two Americans) flee. The Viet Cong take several prisoners, kill others, and 'relieve' the camp of several mortars, rifles machine guns, 105 mm recoilless anti-tank rifles, and—most importantly—significant amounts of ammunition. When the attack begins there is enough time to radio for help. But, by the time planes and helicopters arrive—the Viet Cong have escaped.

Next morning there is a small 'addition.' The captain and a unit travel by car and truck from the main camp previously taken by the Viet Cong, destroy the fortifications, and gather up the dead. They travel by convoy to the main camp but, on the way back, encounter land mines on the road—one car is damaged, and three men (locals serving with the special forces) are killed.

This whole episode obviously increases the Viet Cong's prestige in the area, and the Americans decide to clean it out. Patrols from the 101st Brigade [*sic*] are sent out to 'make war.'

I am writing tonight from a nearby Special Forces camp. Permission has arrived from General Pearson's headquarters for me, tomorrow morning, to join a company of infantrymen currently reconnoitering the area. In this camp as well, they tell me—without going into details—that relations between American 'advisors' and Special Forces command are problematic, making staff changes essential. The entire business is complicated, but typical of the overall situation in Vietnam. At Pearson's headquarters, I review a detailed briefing on unit operations, from the time it arrived in Vietnam. Pearson is very energetic but far less impressive than Larsen. Larsen is (as a man—I cannot evaluate him as a general), the most impressive commander I have met here so far, most especially by virtue of the inner fire that informs his words and facial expressions. Pearson looks tense and tired—although he seems to enjoy the intensity of his unit's activity.

Pearson has studied his brigade's operations in minute detail and promised me written diagrams and other material. After that, we dine with his staff, and from there I travel here by helicopter, after it has become dark.

During dinner, the question is debated: what to do with me? Pearson begins by reminding me about their patrols that go out into the area for an entire week and call for help only when they encounter large forces. However, when he sees that I really want to join such a patrol—he backs down. After extended negotiation, he allows me to join this group for a few days.

While I am receiving a briefing from Pearson—also during dinner, conversation, etc.—I cannot shake off my negative reaction to all the self-aggrandizement that I observe. I want to assume that this is not a personal issue but rather a national trait, and that a foreigner like myself doesn't understand it.

Pearson asks me in which newspapers I would write, and whom I had met of the Army Chiefs. When it becomes clear that my contacts have public relations value, he keeps on emphasizing—after every phrase—that his is the best American unit. I assume that this comes from American reality—perhaps only someone who constantly promotes himself and is praised in public merits advancement and appreciation; perhaps it is simply awareness by the generals of the power of the press.

In any event—although this does not in any way harm prosecution of the war—I cannot overcome my feeling of revulsion at how American generals are so concerned with self-promotion and must think every day of how they appear in the press and worry that news of their exploits might not reach the public. I do not believe that this contributes to the human or military standing of American generals.

The Special Unit camp taken by the Viet Cong has no artillery cover.

It is interesting to know whether conditions might be created whereby farmers do not leave their villages or obtain government support if they become refugees. This would force them to choose between loyalty to the government (inform on the Viet Cong and hand them over)—with shelter and protection—and punishment for capitulation to, and accommodation of, the Viet Cong. It is easier to make them into refugees—but this complicates the situation and delays normalization.

The intelligence sergeant tells me that, during interrogation of the Viet Cong, they sometimes spit in his face and say—'kill us, we are ready, and not afraid to die: rather, it is you who fear us.'

There is no question that some areas here are under complete Viet Cong control. The Americans—and of course the Vietnamese—are afraid to enter these without significant forces.

In addition, every so-called 'cleansed' area, as soon as it is vacated, returns to the way it was before, despite the Viet Cong. Americans rule is limited to where they are physically present.

It looks like there is not a single safe road, open to general traffic, in Vietnam. On some roads, traffic moves by convoy, and others are only used by the locals (on bicycles, because the bridges are blown); they pay a 'tax' to the Viet Cong.

I have been told that in different regions—for example around Duc Kho (*sic*)—the security situation now is worse than it was in 1962; at that time there were government forces in the area, and the Viet Cong did not disturb them.

I think that the general opinion of the sergeants working in intelligence (at least the ones I have met) is that, if the North Vietnamese stop fighting, the Viet Cong will disintegrate into groups of 'bandits,' without outside aid or direction. This, despite the fact that, today, they are more stubborn and much more mobile than the North Vietnamese Army, who are conscripted for 'compulsory service' and not volunteers. The Viet Cong elite are found in the Delta region (Saigon). But sometimes the opposite occurs, and whole Viet Cong units go over to the 'Special Forces.'[20]

Different American commanders have different of approaches to the war. But, examined closely, these differences are small.

In addition to all other slogans, the sign 'soldiers and diplomats' hangs over Pearson's headquarters gate. He explains at length to me that the army's job here is also a public mission: to acquire the people's trust and friendship. In short—a diplomatic task.

The Americans here are winning everything—except the war.[21] Their aim is to end the war and pacify the entire country. By contrast, the Viet Cong's aim is to convince the people that there will be no peace—only war—while the Americans are here. Now—the Americans are in ascendancy, striking and destroying both the Viet Cong and the North. However, all this is far from a real victory. The opening of the roads is emblematic. It is easy to damage traffic. Unless and until free traffic is assured it is difficult to say that the country has quietened down. I believe that Larsen's definition—the aim is to achieve a guaranteed agreement by Ho Chi Minh to stop the war and secure the seventeenth parallel—is more correct. But this will probably not be achievable without taking Hanoi. The objectives here are twofold: stop the war and legalization and documentation of the ceasefire.

The sergeant of intelligence tells me that:

1. A Viet Cong patrol (he doesn't know when) has destroyed an American patrol within a period of 45 minutes.
2. He doesn't believe it possible to find Viet Cong units hiding in the jungle grass, from the air—they wear camouflage caps and walk in zigzag, so that the grass covers them.

It is a fact that the number of Americans volunteering to serve in Vietnam is greater than the number who are accepted (This applies only to soldiers in the standing army who volunteer for Special Forces duty). There are two main reasons for this:

1. They receive tasks with responsibility; here, every sergeant is 'king.' In the United States, by contrast, there are endless training exercises and strict disciplinary supervision.
2. They are professional soldiers, whose profession is war; here they are permitted to 'practice their profession'—learn how to advance and distinguish themselves. 'This may not be a big war, but it's the only one we've got at the moment.' Each one knows the significance of his service here in his personal career and wants to exploit it. From this point of view, it is a fantastic training ground for the American Army. The army would be the last ones who want to relinquish this war (unless, obviously, it is won).

Two things stand out during the flight:

1. Cultivated and non-cultivated plots. Near the houses—or in specific regions of a contiguous area—plots are cultivated. By comparison, there are huge non-cultivated areas.
2. In populated areas (apart from regions near towns), the road network has completely died. Long black (asphalt) arterial roads lie empty of traffic (the same applies mostly to the rivers).

Pearson emphasizes that the French—who fought here—could not activate their airborne radar (without all-weather vision), allowing them to bomb accurately, during the monsoon season. Based upon this, the Viet Cong wrongly think that the same applies for the Americans: it does not, because their B-52s have all-weather radar.

We spend the night of 14.8. at the Special Forces camp in Dong Tre.

The Company on Patrol
15 August 1966

At. 7.00 am, we leave Dong Tre by helicopter, to join the company located 18 kilometers from here. The company must return here on foot, patrolling at the same time.

I sleep very little. My escort—Colonel McCord—is a devout Catholic but has never been to Israel. However, he knows it well, from the Old Testament. Hours after I finish writing and turn the light off, he is still asking me detailed questions about Israel and what is happening there, with amazing knowledge of its geography, and the main issues with we must deal: the bends of the River Jordan, irrigation in the Negev, immigrant absorption, relations with our neighbors, phosphates, potash, etc. He is particularly enthused when citing biblical names such as Beer Sheba, the Sea of Galilee, etc.

During flight signs of heavy B-52 bombing can be seen. Huge craters create bare patches in the jungle, and rock falls on the steep slopes, which crumble easily. There is no doubt that the planes aim accurately—bombs are dropped by radar, and reach their target rapidly, when called in. However, men must be kept about 3 kilometers from the bombing site, so these planes can only be used when there are no soldiers on-site. It takes a long time to march 3 kilometers under difficult jungle conditions (according to Colonel Emerson[22]), so in that case it is better to use ground artillery, or (if weather conditions permit), nearby spotter planes, without necessity of heavy bombing. On the way, I tell Colonel Emerson that, during our delay at his company headquarters (2nd Company of the 502nd Infantry Brigade), everyone speaks highly of him. The day after tomorrow, he will end his year-long tour of duty here and return to the United States. During his term of service here, he has studied the issue and developed a theory on how to conduct guerilla warfare. His doctrine—at least in part—depends on splitting up into small units. He calls this a'checkerboard'—because he divides the area into squares (obviously non-schematic). He appoints each subunit (a platoon, maybe less) to scour a specific area on the 'board' (which might be 12 x 16 by 2 x 2 kilometers for each platoon). During the afternoon, I meet with him in his battalion headquarters: over a cold beer, he enthusiastically explains the main points of his doctrine, giving me (a great deal of) material in writing. Apparently this battalion has carried out several successful—even very successful—operations. They recently ambushed and destroyed ten to twenty North Vietnamese soldiers in each of several operations. His men lie in ambush during the morning (6.00–9.00 am) and evening (5.00–9.00 pm) hours. During these times, the Viet Cong come down from their hiding places into the valley roads, to fetch water or move from place to place; they assume that there no Americans in the area, or else that they have already settled down for the night.

This is obviously only possible if the unit is in the area for several consecutive days, with no continuous helicopter communication. The 502nd Regiment (or 101st Division) has the unique quality of arranging patrols and stationary ambushes lasting several (up to 7) days, without reinforcements or extra supplies. The unit is split into small subunits: even an augmented squad, certainly a platoon. This method is not accepted in other units such as the Marines of the'cavalry.'

However, the battalion has had no luck for the past 2 days. The Viet Cong have used individual snipers and caused several casualties in the battalion. Emerson tells me that he would have sent the small units outside artillery cover, but Westmoreland's orders were to operate only within artillery range (except for Special Forces camps).

We reach the company at 8.00 am. The helicopter carrying us returns. McCord and I join the company. During the first 2 hours of walking (with the company), no signs of the enemy can be seen. After that, we approach an area in which they are apparently active. At first, only bamboo darts embedded in the road are visible. Anyone walking barefoot on these would suffer serious harm. But high American army boots easily trample them down; this 'Maginot'[23] method looks wretched and pitiful. After that, we encounter barbed wire fences that block the road, also mines. A buried leopard is the first sign of mines. A fresh grave and a leopard's tail (revealed by the flooding rain) sticking out, can be seen on the side of the road. A few steps behind is a pit containing an (American) hand grenade from which the firing pin has been removed. Anyone stepping on the leaves covering it would make it explode. There are four such grenade-containing pits; others are empty (but up to a meter deep), covered with thin branches and dust. Anyone walking at night would certainly not see them, and fall.

The mines are cleared and exploded, and we continue on. (I assume that the leopard trod on one and was killed, and the Viet Cong buried it, simply to stop it from stinking and revealing the mine).

After another 2 hours of walking, we approach our objective: Dong Tre. We must pass a populated plateau, on whose edge is a stream, dividing the plain from the army camp. I have a chance to investigate one of the houses scattered in the plain and see the 'heavy industry:' sugar ground from sugar cane. (Water buffalo turn a central pivot—like olive oil production—and two round tree trunks, one on top of the other, crush the sugar cane placed between them. After that the juice is cooked, and sugar crystallized from the thick fluid).

When I return to the unit, I hear shots. The unit is scattered—three platoons and several snipers walk in line as rearguards (afterwards, I am told that five men dressed in black pajamas were seen). They let the unit cross, to attack it in the rear—with their own snipers. They do not miss. The first bullet, with a relatively short range—from the hill through which the road that enters the valley passed—hits the lieutenant who was walking with the rearguard. Even worse—the bullet hits the middle of the grenade hanging in the officer's belt. The grenade explodes, and the young man is killed instantly—his legs are smashed, and his intestines spill out of his belly. The company begins to maneuver right, and left, to outflank the shooters, and covers the hill with machine gun and mortar fire. After that, the artillery lands, but I doubt whether any of the snipers are hit. They have enough time to jump down and escape, after they have shot about twenty bullets at short range and a couple of dozen more at a longer range. The company has only one wounded—but the appearance of the mangled body and the feeling that 'the Vietcong have 'put one over on us' leaves us with a bitter taste in our mouths.

A helicopter arrives and takes the body back. The artillery fire lasts for a while—and the fighting ends.

Clearly the Viet Cong can cause damage with individual snipers, if they can make the enemy out at a distance. Their best chance—is when a single file of Americans moves around in the area. However, if they concentrate a larger unit and wish to engage in a real battle—American firepower, mobility and communication facilities turn the Viet Cong attack into defense and defeat. This American unit has also killed dozens of North Vietnamese soldiers by ambush, when they land and wait in ambush, while the North Vietnamese move on the road and walk into the trap.

With the Commanders

I return from the camp to Emerson's base to talk to him again, and after that, to the brigade. I hardly have time to wash myself a bit when General Pearson arrives and suggests that I join him on an inspection of his artillery unit which have clashed with the North Vietnamese earlier today.

'Why not come with me?' Do I have a reason not to?—I am already dead on my feet from exhaustion. But I get dressed and we travel by helicopter. By the time we arrive, it is getting dark. We find out what happened. A platoon (from a company of riflemen guarding the artillery unit) was patrolling in the vicinity. On the way back—about 500 meters from the artillery unit—they encountered two North Vietnamese soldiers, sitting in ambush—or else maybe just resting—behind a rock.

The first American soldier that arrived at the spot did not see them and jumped down from the rock. His comrades heard a shot and thought that a bullet had been discharged while he was descending. Two other soldiers followed him: while they were jumping down, a volley was shot at them from 3 meters away. One was killed; the second was hit, and the bullet scraped his rifle. He returned fire and killed both North Vietnamese. During the commotion, a grenade exploded and wounded someone else. In total, two North Vietnamese soldiers are killed: the Americans have two killed, and one wounded.

While we talk to the young soldier who killed the two North Vietnamese, a Red Cross [medevac] helicopter arrives and evacuates the dead and wounded.[24]

A luckless day. Sometimes this happens.

While flying at night and talking to Pearson about what was happening—I cannot get away from memories of night flying in Israel during battle. Here the country itself, with its civilian inhabitants, is of no interest to the (American) commanders. By contrast, in Israel, our main concern is for the settlements beneath us, some dark, others dimly lit.

While walking in the jungle, I search for fruit, so see whether it is possible to 'live off the land.' I find a bunch of green bananas (they can be picked, and ripen, after a few days in a knapsack). A few small rock-hard guavas, also a reddish, oblong fruit—in an emergency I could certainly have chewed on them. But I could not have lived off what I found.

According to the company commander and Emerson, the North Vietnamese soldiers they last encountered (fifteen were taken prisoner 2 days ago), looked defeated. They were suffering from malaria, malnutrition, and the heavy aerial and artillery bombing was shaking them to their core.

A kind of aggressive red ant found here makes a type of closed container, with leaves tied one on top of the other by sticky mucus. They lay their eggs in this nest and multiply. When I see this, this morning, I understand why I had been bitten so badly during my night patrol with the 'cavalry'—I must have leant on a 'cluster' of such ants.

North Vietnamese snipers do not influence the unit's spirit. Those that die, die, and those left alive, go on living—tomorrow morning they will again be ready to search for any signs of the North Vietnamese in the mountains—in full knowledge that if the enemy will find them first, the initial part of the battle will go badly for them.

During patrols, meals, and other periods of rest, I examine the officers, for, amongst other things, their motives to volunteer for duty here. Some refer to the ideological mission of a war against communism. But the main (and perhaps the only) reason is that they are professional soldiers. War is their job, and here they can practice their profession, learn it, and advance in it. Their service here—or its lack—is noted down by their superiors when promotion time comes around. But even the latter is a secondary reason. Their main motivation is simply a wish to 'practice their profession' and excel in it—they want this whole-heartedly. Obviously, I am only speaking of officers and non-commissioned officers in the permanent forces, not privates.

When I compare this spirit with the aphorism that 'the American army will get tired'—the saying seems like a naïve joke. The American military—singularly and collectively—celebrates every day of war here. It is an outstanding opportunity for them to train their soldiers, and their replacement after only a year of service is no coincidence—to give as many soldiers as possible the opportunity to 'take the course in Vietnam.' They are trying out new kinds of weapons here—especially, in the air, artillery and infantry (almost without armored support vehicles), conventional weapons, electronic communications equipment, radar, night detection, etc. The American people and politicians might perhaps conclude that this war must be ended, but I have no doubt that the American Army is extremely interested in its maintenance and continuation.

Time after time, conversations return to the difference between the current American, and the previous French, way of prosecuting the war. American advantages include:

1. Their weapons, and the quality of their army, are superior.

2. They are waging war here whole-heartedly and with (professional, not ideological) enthusiasm.
3. They want the war to continue.
4. As a country—they are not here to economically exploit the Vietnamese people and their country, but rather to strengthen their political and strategic standing in the world. This does not damage Vietnamese individuality (as said by those with no political enlightenment), Just the opposite: they believe that American influence here is mostly good (dollars and work).
5. By contrast—the French were more accepted here as individuals. They were here for 100 years, their culture penetrated the populace, and many locals still speak French. The French are generally more comfortable bringing peoples from different countries together (objects of disputes, etc.).

Even after these conversations with Colonel Emerson, I still believe that all this talking about different approaches to war needs clarification. There is certainly place for strategic and political alternatives (conquering Hanoi, Cambodia and Laos, increase in the number of American soldiers, use of nuclear weapons, concentration of attacks versus splitting up, etc.). But differences in fighting tactics are small:

1. Variation in personality of different commanders (greater or lesser aggressiveness, traditionalism, risk-taking, etc.
2. Organizational variation, for example the first 'cavalry' division has an abundance of helicopters and thus are very mobile.
3. Nature of the terrain: plains, mountains, populated areas, etc.
4. Seasons of the year, like monsoon, etc.

5. Different political considerations: even if there was strict guidance from the Pentagon about ways of conducting the war there would still be certain tactical operational differences. But these would not constitute different 'philosophies,' or solutions to the problems of this war.

Emerson believes that the North Vietnamese Army has three advantages over the Americans:

1. Superior intelligence—they are expert spies and constantly track the Americans. They are helped by the population both outside, and inside the American army camps.
2. They are experts in splitting their units up, the better to infiltrate the jungle and concentrate their forces prior to an attack. This allows them to exploit the individual ability of their commanders. This splitting up places responsibility on the shoulders of many low-rank commanders, and is especially important in such an isolated region with thick forests, etc. It allows infiltration without detection, and rapid activation (in the circumscribed area) of fire from each squad or platoon (they do not move in the area with more than fifteen men).
3. Control of a soldier's motivation; there is instruction on the commissar system and political indoctrination. The Americans are complete ignoramuses in this area—in contrast to the soldiers of North Vietnam.

Emerson tells me that he once captured fifteen North Vietnamese, and that he 'tipped his hat to them': with the conditions under which they fought, they showed such courage that he had to literally 'blast them' out of their trenches. (The Americans use a specific gas for this purpose—non-poisonous, but it burns and terrifies).

Although the Americans are destroying many North Vietnamese units, they cannot convince them to surrender. On the other hand, in Korea large units surrendered. Emerson explains that as an error in American psychology. Northerners are convinced that, should they fall into American hands, they will be taken out and killed. That is one of the reasons why they refuse to surrender, even when strikes are heavy, and their situation is hopeless. Emerson is not free of self-aggrandizement (not as bad as General Pearson, who is a 'world-class' boaster).

There is a great 'game' played here with names, slogans and formulas; they are really different headlines for the same thing. Westmoreland demands 'improvised patrols'—not only patrols for reconnaissance or battle, but anything else that turns up—hence the expression 'recondo'—combination of reconnaissance and intimidation, in other words a kind of 'Ranger' commando. Pearson unashamedly calls his unit's activity 'the odyssey of the 101st—the finest unit in the world.'

Emerson has had a recent successful operation: his unit killed 300 (of 350) men from a North Vietnamese unit, compared with six killed and forty wounded in his own unit.

16 August 1966

I am still at Pearson's headquarters. This morning, he gives me material on their large operation (according to him the most successful in Vietnam to date), with sketches and explanations.

Yesterday, amongst other things, Pearson told me that he has a 'starlight device' (a kind of binocular, which doesn't depend on infrared rays), through which a man can be seen in the dark at 300 meters. With the 'checkerboard' search method, searching is only done with small units; however, if one of these units encounters forces in significant strength, remaining units are concentrated—or else units are added, to surround the enemy, block their retreat, etc. Then, of course, the (regiment) commander takes over, to coordinate the unit's activity and supervise it. (They now call Emerson 'Mr. Responsibility.' I have no idea where this name came from. I am 'responsible' for everything—the area, the situation, and the units under my command—everything. And I had naïvely thought that I command but am not 'responsible'!).

His theory against control concentration is directed at regimental commanders, who tell every squad what to do. With today's communication facilities, Westmoreland can speak to each platoon commander and give him orders. 'Believe me, the top brass do just that ...'

For this reason, he argues for splitting up of a large unit. The theory of force concentration, etc. was perhaps correct during the Second World War, when the exact enemy position and extent of its forces were clear: this made concentration of large forces for attack, under a unified command, possible. But this does not work here; its real meaning is—unnecessary and unjustified interference from on high with subordinate authority, leading to premature force concentration—before it is clear where, and how strong, the enemy is.

This morning (16.8) I go out to see one of the battalion's training exercises. The squad performs better than the platoon, which doesn't do so well. The squad is made up of two equal parts of five and five, with machine gun, grenade launchers (50 millimeters)[25], etc. Each company has its own sniper. Exercises are done with live ammunition: several days ago, twelve men from this unit were wounded in this way—today, only two. First aid is applied (The medic works nimbly, without stopping chewing gum for a single instant), and they are sent for treatment by helicopter. I think that, in view of the upcoming jungle operations, it is justified, with untested beginners, to use live ammunition during training. In this way, their lives may be saved by simulating exact conditions. They must not be permitted to be surprised by the jungle, or by the fighting that awaits them.

After the exercise, all remaining non-commissioned and commissioned battalion officers who have been here for a while, assemble (service in Vietnam only lasts for a year). We talk—they answer by questions about the war, describing battles in which they have fought, and who initiated them; were the Viet Cong able to escape battle, or did they find themselves in a position where there was no alternative to fighting? Change (improvement) of the war situation during this past year, etc.

They tell me that the situation has improved greatly. A few months ago, they were cooped up in camp, on the defense and fearing a Viet Cong attack. Now they move around the area both day and night, searching for the Viet Cong. The enemy avoids encounters with the Americans, but, despite this, the Viet Cong believe that they will gain the upper hand—when they succeed in luring a small American unit into an ambush, etc.

On the question of prisoners—the battles are so fierce, and so much fire is directed at the Viet Cong's positions, that even if they wish to surrender, they cannot even raise their heads, talk, put their hands up, or come out with a white flag. They have only two alternatives: to be killed or try and escape to their jungle trails in the rear. There is a high percentage of Black non-commissioned officers (sergeants, etc.) here. They volunteer especially for this unit—because of the higher pay. This difference in dollars per month plays an important role for Black soldiers.

While flying from place to place, we pass the coast (the exercises also take place right on the coast). The exceptional beauty of the jungle-covered mountains, bathed by clear, blue water, is strikingly impressive. Route no. 1 twists its way through part of the vista. Of course it is not in use now, but in the past, this was the central road from the North to the South. In the valleys—rice can be seen; plowed plots in preparation for the coming season (planting in September). Green plots, other plots covered with water—that look like large glass squares from the air—the clear green of the rice is beautiful, clear and refreshing, compared with the dark green of the jungle.

In the jungle, there are large sectors (entire mountains) of trees bare of leaves, with grey-brown branches. This is the result of defoliant spraying,[26] which has been done to see what is happening in the forest, both from the air, and on the ground.

The brigade commander escorting me speaks positively about the spraying. But I doubt its usefulness. Instead of being green, the undergrowth is yellow, and branches—even bare of leaves—cover the area. Additionally, new growth has already appeared, bushes and grass that were not damaged by the spraying—or which overcame it. The jungle is stronger than all instruments of war, fires, bombings and sprayings. A few monsoon floods wash away—with water or soil erosion—all traces of destruction. New growth sprouts up powerfully, and plants become intertwined to the point of saturation—when the trees and bushes don't leave any space for any more bamboo to grow. The only one capable of overcoming the jungle and being able to work with it, is the Vietnamese farmer. Short and lean, he cuts, morning after morning, grass taller than himself with his machete, uprooting grass and roots with a farmer's stubborn persistence. Only he and his frail-looking wife, bent over next to him under a straw mushroom hat, planting the tender rice seedlings with faithful hand—they are the only ones from whom the jungle retreats. Only they know how to turn the rotting sediment of jungle waste into a blessing, to line up and fertilize the rice beds. Only they, with the accumulated wisdom of countless generations, know how to change this wild, cruel undergrowth into protecting living walls, restoring wind and shade to the tiny plots, so that even the tenderest shoots can grow to maturity.

Yesterday, during flight, Emerson speaks with his platoon commander (perhaps the one that remained in the area). American style, he tells them when they part (he returns to the United States tomorrow)—'you are the finest platoon in the American army, but don't quote me on that'—all in complete seriousness. America, America.

The jungle has taught Giap (the North Vietnamese Army Commander)[27] the correct mobility tactics. He is also teaching others—Emerson, 'Special Forces,' etc.—that the only way to advance in it is in small groups, on separate, parallel paths.

In each location, there is a clear line between cultivated and abandoned areas. Most Viet Cong areas are abandoned because of the fighting there. It is not enough that the farmers flee from the bombing and hardship brought about by the war—but the government sees them as Viet

Cong collaborators (who collect rice taxes) and treats them as hostiles. They can do nothing else but leave the area.

The government takes all the rice (from territories under their control) and stores it in a large warehouse (in this region—30,000 tons). It notes exactly how much rice it has received from each farmer and lets him take back home the amounts necessary to feed his family and also sell—all under supervision. This is done to prevent the population from giving rice to the Viet Cong. Maybe I'm wrong, but I don't believe that such a bureaucratic method will work. I have been told that this method was successful last year.

In yesterday's newspaper, Prime Minister Ky[28] wrote: 'if the situation continues to improve as it has during the past 14 months—I will ask all foreign forces to leave Vietnam.' (This, in reaction to the voices who have said that the war will continue for another 15–20 years. Tomorrow another newspaper will appear, and someone else will say something new).

During the afternoon, I visit the artillery units again. The Special Forces who were there yesterday have been changed. The new unit looks much more invigorated than the previous one, who sat yesterday helplessly, wretched, reminding me of the *ghafirs*[29] (home-guardsmen) of 1936. The artillery pieces are badly positioned from a tactical point of view: they are surrounded by mountains and can be approached undetected and attacked from the mountains. It was apparently chosen because of the rivulet that passes through this area; most of the soldiers remain here all day.

This morning, we receive a request for artillery fire from a helicopter (OH-13) that was traversing the area. Three Viet Cong (with black pajamas) had opened fire on it. The artillery roared, but the location was misidentified and the guns 'corrected' to add an additional 2 kilometer range. However, by the time the shells arrived at the right spot, the Viet Cong had escaped and hidden themselves.

The Americans talk of their control here in terms of percentages—60% of the population living in the central coastal region. However, from a military point of view, this really means that 80% of the area is outside their area of control (even if they enter it whenever they wish, for military operations). This area can serve as a Viet Cong base—for organization and night lodging.

Problems created by entrance of the Americans into large areas are:

1. Evacuation of the local population, turning them into refugees;
2. 'Americanization' of the area;
3. Creation of conditions where only the existence of American forces in the area creates security.

It is difficult to withdraw from such a situation: when the Americans leave—the Viet Cong are bound to return, something that the population regard as natural; in other words, the Viet Cong are their alternative. The only other way is to transfer the area, as quickly as possible, to South Vietnamese army and police control. In this way, the people will see the Vietnamese Army as their alternative to the Viet Cong.

17 August 1966 Tuy Hoa
General Pearson's headquarters

Today is my last day here. In the afternoon I return to General Larsen's headquarters.

I go out to observe a local South Vietnamese battalion—three companies (about 450 men). The area was bombed by planes and shelled by artillery, only then were the companies landed, like an industrial conveyor belt—four helicopters landing every few seconds—the soldiers (five to six per helicopter) jumped out, and the helicopter took off again. Twenty jumped down at the same time. The helicopter unit serving the battalion (16) made two rounds, to drop off the entire battalion.

The company did not come under fire, but they noticed a number of suspicious men in the area and opened fire on them. The area where the battalion landed was cultivated, and under Viet Cong control; until now, the population has not fled. But now the war has reached then—they certainly will not be able to remain here.

The battalion will be in the area for 5 days and comb its assigned 'square' (15 x 15 kilometers) by advancing from north to south. The area is divided amongst the companies. Each company must move from one controlled (mountain) area to the other during a period of one day. If they encounter the enemy, the company will concentrate, and attack. Throughout their operation, the company will remain within artillery range. Each company has several American 'advisors' and they, of course, are the ones who decide and maintain communication with artillery and air force command.

During the night, two convoys land on mines on the coastal road, near our camp. The road had not been used for the past 3 weeks, and in the meanwhile, mines had been laid.

While we are in the air (we watch the entire operation of landing the Vietnamese company from the air), I hear the commander of the battalion (which is still in the rear) asking the helicopter if the company that already landed had encountered the enemy. The pilot replies: 'I haven't heard a word from them, but I see (from the air) a shadow near the men—a sign that they are standing. If they had encountered fire, they would not be standing up straight.'

Again—advantages of the American Army over the French:

1. Helicopters: free of road restrictions;
2. B-52s (radar-directed bombing);
3. Limitless means (money, artillery, manpower);
4. Washington is behind them, and there is mandatory conscription;
5. They have no other front line (Algeria);
6. The Army wants to fight, is enjoying this war and prosecuting it in comfort (provisions, not too much effort, one year service

with leave) and unlimited equipment. A huge feeling of superiority over the Viet Cong.

Determination of the landing area is done with prior low-level air reconnaissance. They are now taking photographs, hoping that these will be sufficient to prevent early warning by the enemy because of reconnaissance that precedes landing.

The Americans have missed, and not exploited, this year's main surprise opportunity—ability to fight during the monsoon season. If they had left the Northern Army to its error, and allowed them to initiate the attack themselves—they could have been victorious.

After the South Vietnamese units land, we fly off to observe the landing of a battalion from the first 'cavalry' infantry division. The first part, transport—is very good: the unit is rapidly deployed after landing. However, the choice of the place to land, and therefore the company's starting position—is very bad. They consider ease of landing more than tactical importance of the terrain itself, and do not grasp the significance of the heights. Apparently, the battalion commander understands this, because at least one company is immediately transported to another much higher location.

Transport of the guns is not centralized, but this is not important. Preparatory firing (air, artillery, transfer of guns) is not done according to information, but in routine fashion.

Seen from the air, zigzag dug canals are clearly American or South Vietnamese. The North Vietnamese cover their positions with camouflage.

The young man who killed the two Viet Cong soldiers—after being wounded himself—is externally unimpressive. He has the face of a messenger boy in a grocery store. He doesn't raise his voice while speaking, and it seems to me that he would have felt more comfortable saying that someone else had done the deed.

At 2.00 pm, I take leave of General Pearson. He gives me a regimental souvenir, we have photographs taken together, and I invite him to Israel. Beyond formalities, we say goodbye in cordial fashion. When one sleeps in adjacent tents, uses the same shower and latrine, and scampers around together in a helicopter for several days—one connects, to an extent.

An old acquaintance, Major Castilia, is waiting for me at Nha Trang airport. I have nothing specific to do this evening (a few days ago, General Larsen informs me that he would not be here this evening). In the morning, I will fly to Saigon.

Break in Saigon
18 August 1966

This morning, we walk around a bit in Nha Trang, to see the town and harbor. The harbor is small—I don't understand why, under current conditions, the Americans have not developed sea transportation along the entire coast. Most of the cities are coastal, or near large rivers. In the absence of a main road (bombed, etc.), large-scale coastal transportation could be established.

At noon, when we arrive at the (Saigon) airport, I see that the facilities there have been disrupted. Last night, the Viet Cong shelled the airport with mortars. I cannot stand properly on the damage, and the telephone doesn't work. My plane drops me off in some remote corner or other. With no other alternative, I sling my rucksack over my shoulder and set out on foot. Eventually, I reach a kind of taxi and arrive at the hotel in it.

I have lunch at Barry Zorthian's[30] home—he oversees the American Information Service (he is an Armenian). His assistant and political advisor (of Lebanese extraction) is very pleasant. Two other Americans join us for lunch. Ambassador Porter[31] was once appointed to the Middle East and served as Consul-General in Jerusalem. I don't know the fifth person. The conversation topic is, quite obviously, my Vietnam impressions. Each one of my criticisms points leads to an argument, which becomes very heated at times.

At 4.00 pm, I meet with the Ambassador—Henry Cabot Lodge. He carefully gauges my impressions and asks me why Israel refuses to send instructors here; if I had any problems, I should consult him. To facilitate this, two more men enter the room, and give me their addresses.

At 6.00 pm, I meet again with Zorthian and his assistants, to arrange the program for the last part of my visit. They take it upon themselves to explain the position here, and its different causes. My program will be complete in a few days' time, but in any event, I will be here for about 10 more days. I will begin my visit with Welfare Administration [*sic*][32] this coming Wednesday.

In the evening, I meet with one of the Americans from Lodge's office. At his initiative, he gives me a 'briefing' on the current political situation. I am not terribly surprised. He 'licks his chops' at America's successes—even though he cannot name a single European country—except for Spain—which unconditionally supports the United States.

While he is speaking about the importance of the United States 'standing by its word' and helping countries to which it had given guarantees, I start thinking: maybe there is something to this—but in the opposite direction. With the war in Vietnam, the United States has unilaterally established its status as international policeman, with permission to interfere with any conflict it wishes.

The way he sees it, the North will tire of the war—they will not show any sign of this before the American elections (November), because they are depending on differences in American opinion on the war.

The rest of what he says is nothing but repeated platitudes. But it is interesting how he emphasizes that the only things the United States knows about North Vietnam come from:

1. Photographs;
2. Decoded radio transcripts;
3. Interrogation of low-rank prisoners. In the same way, he confirms that the North's ground-to-air missiles do not work.[33]

19 August 1966
Saigon

I have lunch with John Ridley, a representative of *The Daily Telegraph*[34] here. He has received a telegram from his newspaper asking him to assist me in any way he can.

I tell him that I have a Burmese visa, but that I do not intend to visit at a time when Ne Win[35] is not there. He urges me to change my plans: not one single British journalist has been successful in obtaining a Burmese visa, and I would be able to meet prominent personalities there, and hear and interpret their thoughts about the future of that Asian area—I should do it, even at the price of giving up a trip to Japan, which has in any case been 'covered thoroughly' by the media. I will try to follow his advice. He questions me closely about my impressions. He shares the premise that America is simply 'thumbing its nose at the world,' and doesn't care about the opinions of foreigners. (I tell him that they are more interested in the political opinion of Jewish women from Brooklyn—because they can vote in American elections—than those of Bertrand Russell[36] in Britain).[37] He is very critical of Americans, and doubts whether they are really making progress in understanding their war aims. Like most journalists here, he is more interested in the political than the military-technical part of the war—even though he was the *Telegraph*'s military correspondent during the Korean War.

During dinner (Hotel Caravelle) we meet up with the British military attaché here (he has already been here for 8 years!). He invites me for lunch at the embassy tomorrow. If I don't travel outside the city—I will come.

The morning newspaper carries a 'bombshell': Bertrand Russell says that the Russians have told him that—as soon as Ho Chi Minh requests it—they will come to protect the skies over North Vietnam. This report has been trumpeted in local newspapers (I don't believe that the Russians would entrust such an important announcement to Lord Russell).

The local press has become distinctly chilly towards Americans in recent days. One of the editorials was very critical of the Americans increasing the number of soldiers here—without having taken counsel with the South Vietnamese government. In the same way, the Prime Minister, General Ky, has let it be known that, if conditions continue to improve, he is liable, in two years' time, to ask all foreign troops to leave Vietnamese soil. Westmoreland's reaction:

1. General Ky's words have been distorted; what he really said was that he would request only part of the army to leave.
2. He hopes that the General would 'think twice' before he asked the Americans to leave.

In the evening, I meet with my cameraman from United Press. He tells me of the unnecessary bombings of populated villages. A picture of horror—all the people killed by the bombings. Those who told him about this—were Americans. I also saw them bomb a village that I thought to be inhabited—but I only saw it from the air (during the last landing of South Vietnamese troops).

One of the important questions is whether the Russian SAMs really have failed—if this is so, it will be a hard blow for Russian prestige.

20 August 1966
Saigon

I have lunch with the British military attaché. Before lunch, in his office, (map room), he explains the current war situation, and its development over the past 2 years. His main point: Any exact definition of regions controlled by Americans or Viet Cong is totally unrealistic. There is no region which the Americans cannot enter; but the opposite is true as well:—the Viet Cong are also active in every region. Apart from that—'control,' only lasts while Americans are present. The minute they leave—the Viet Cong returns.

Our main difference in opinion is as follows: He keeps on talking about South Vietnam, explaining that the war is about control, security, and influence—My opinion is, that the Americans are currently occupied fighting against, and want to defeat, North Vietnam, justifying the merit of their presence in the South; the same applies to recognition of the North from the seventeenth parallel. For this purpose, the Americans are trying to destroy and crush the army of North Vietnam and kill all its soldiers who reach the South. If their goal with the North succeeds—the problem in the South will be solved of its own accord. If not—any imposition of order on the South is temporary and unstable.

We both agree on one thing—all talk of areas and percentages of control, etc.—is totally meaningless. If there is any progress in the American situation here, it is with accumulating pressure being brought to bear on North Vietnam. Their losses, and those of the Viet Cong, are increasing, and both have been forced to split up into small units, capable of terror but not of battle against regular forces (if this really is an achievement on the part of the Americans).

The British attaché gives me a letter of introduction to the British attaché in Bangkok, to facilitate my meeting with the exiled Vietnamese General Minh.[38]

General DePuy's Division[39]
21 August 1966
Lai Khe[40]

At 10.30 am, I arrive at the 1st Infantry Division under the command of Major General DePuy. I am told that the Pentagon believes him to be their most talented general here. He tells me that, during his three years here, he has been sent several times by the Pentagon to 'present the Vietnam problem.' He is a pleasant, simple-mannered, quiet-spoken man, who knows what he is talking about. During the evening briefing, one of the officers says about someone that, if they send any more planes, they are liable to be squandered, etc. DePuy glares at him and says quietly—'And I, God forbid, am liable to lose a battalion.' The officer, a colonel, shrinks and says—'yes, sir.' End of argument.

We tour the area by helicopter. Before that—as soon as we arrive—we (both the general visiting from the Pentagon and I), are handed a review of the division's activities during its time here. While flying with DePuy, touring areas of battalion operation, we talk of many things. During the evening briefing, he introduces me to his officers and speaks highly of me.

DePuy believes (with more than 50% certainty) that one of his battalions has been attacked during the night and is preparing reinforcements. I ask—and he promises—that he allow me to join the reinforcements, if they are sent out (he says that he would also go, and I would accompany him). His assistants then enter, to make arrangements in case we go out. In the meanwhile—the roar of uninterrupted cannon fire reflects aid for the battalion on the ground providing intelligence of movement in the vicinity.

DePuy doesn't believe that there is any chance that the war here will end soon. Two weeks ago he met with Nixon, who in his turn had met de Gaulle's representative in Hanoi[41]: he heard from de Gaulle that Ho Chi Minh and the Chinese were prepared to continue the war in the present form:

1. They can hold out (North Vietnam).
2. They assume that the South Vietnamese government will not last.
3. They assume American public opinion to be anti-war.
4. They see this war as a way to drive a wedge between the United States and the Soviet Union.

At the end of the Hanoi meeting, the commander of the North, General Giap, had spoken about the change in tactics against the

Americans. DePuy assumes that the guerilla war will broaden, and large clashes diminish. He maintains that 90% of clashes are instigated by the Viet Cong, when they think they have a chance to win. Usually, if the Viet Cong do not want to fight they do not do so. The jungle provides them with a permanent safe haven.

DePuy believes the Viet Cong to be superior to the army of the North from the point of view of motivational intensity. They know the area and are, all things considered, excellent soldiers. He says to his soldiers: 'In this war, the two best armies in the world are fighting each other (the Americans and the Viet Cong): if we were to fight with equal means, there would be very little difference between us. Our superior firepower gives us an 8:1 advantage. Therefore, we must exploit this as much as our artillery allows—regard yourselves as millionaires (do not stint) with firepower'

There is guerilla activity in some areas—others have less. In active areas, 70% of the losses in DePuy's divisions are caused by guerilla activity. The guerillas fight during the daylight hours and go home in the evening to eat and sleep with their wives; after midnight, they go back to work.

The Viet Cong are very well organized in this region: exact registration, typewriters and hospitals. DePuy sometimes thinks that medical treatment their soldiers obtain in the field is better than American first aid. Despite having regimental bases, the Viet Cong bivouac in company strength. They dig in and camouflage with hats and straw buildings. They have bases and roads in the area. Captured soldiers testify that battalion morale is not uniform. Usually, the spirit of the Communist Party section is strong. It is less so with others. There is very little desertion. A prisoner described his training sessions—a great deal of time is given to indoctrination.

In DePuy's eyes, this war is like a competition between two horses: the American war horse must reach and overtake the opposition 'horse' of the Vietnamese people to the American presence here. This opposition is growing and spreading, and conditions can only deteriorate. First and foremost, because of the difficulties caused by the war, the Americans must end it before the 'opposition horse' overtakes it.

He sees no way to protect civilian roads. To properly secure these, many soldiers are needed, but only a few guerillas are necessary to sabotage a road.

There is a current weakening in Viet Cong operations, caused, DePuy believes, by three factors:

1. Viet Cong have experienced military defeats.
2. The month of August is not war-friendly: streams rise and overflow, there is rain and mud.
3. In order to concentrate on sabotage during the September elections, they are transferring many of their men to Saigon, etc.

The principal American difficulty during battle is difficulty of maneuver in a jungle environment, and problems in:

1. Maneuvering to surround the enemy;
2. Rapidly moving from the area where men are being killed (covered with enemy fire).

They can only control any area when they are physically present in it. It took an entire battalion to open this road (Route 13) to traffic. Afterwards the same battalion receives another assignment—and the road is closed again (bombs and mines). Now, only bicycles traverse it (undoubtedly adding to Viet Cong communications).

General DePuy attaches special importance to their patrol unit (divisional 'squadron,' patrol in depth—they volunteer, when necessary, with additional firepower, etc. They go out for 4–5 days and determine a pick-up area (different from the one where they landed). Their assignment is—reconnaissance, and ambush. They only use radio communication when essential. The Australians—considered professional jungle fighters—go out for 2 weeks without additional supplies: they have special patrol rations. They eat only one meal a day, walk for 5 minutes and then sit and listen for another 5 minutes, and avoid roads.[42]

Infantry companies also go on patrol—squads (three to four per company) sometimes go outside artillery range (knowing General Westmoreland's order not to do so—'it doesn't matter').

The trouble with the American Army here today is—additional helicopters. The speed of airplane manufacture and pilot training does not keep up with the need.

When we visit the area, I am surprised to see how solid fortifications are. The units only arrived 24 hours ago and are leaving tomorrow—but already everything is dug out and camouflaged, with camouflage netting on wooden poles (cut on the spot or brought in). The fighting capacity of the Viet Cong cannot be taken lightly: American perimeters have already been penetrated twice. The perimeter is surrounded with barbed wire and centrally operated mines. Outside, at 100 meters, there are observation posts manned by one or two soldiers. As soon as an American battalion arrives at a new location, the Viet Cong immediately come out to spy, and 'note' the position down. The division patrols (who arrived yesterday) have already clashed with two Viet Cong patrols—and killed four of them.

Although Australians are 'professionals,' they are not immune from attack. A few days ago, two platoons in the area were walking ahead, and one behind (a total of one company). One platoon was ambushed, and *hors de combat.* The rest fought until almost out of ammunition. By chance, an armored convoy arrived in the area and disposed of the Viet Cong (the rain and bad visibility made assistance from the air, and landing of ammunition etc., impossible).

Substantial portions of the area are cultivated. Groves of rubber trees and coffee plants. Right in the middle of the tumult of firing and the noise of planes and artillery, women plant rice, and grey water buffalo roam calmly around.

DePuy relates how the Viet Cong have learned, in their own areas, to live (even in civilian villages) without towns, roads and communications. They will be able to live like this in the future as well: they have returned to their primitive state. He assumes life is the same in North Vietnam—without electricity, etc.

After having tried all kinds of flight methods, they have concluded that flying helicopters in line, one behind the other, is best.

At least one counter-attack unit is always ready—with helicopters.

After dinner (conversation about the Dead Sea scrolls, Sinai campaign, etc.)—DePuy invites me to join his daily briefing. About thirty officers of higher and lower rank go up to the map one by one and give their reports, preparations for tonight and tomorrow.

DePuy is a little tense. For some reason, he is worried that one of his battalions will be attacked tonight. He has received information that some villagers have been conscripted to transport 51mm mortars by road to the unit's vicinity.

In the middle of the briefing, he gives orders to prepare for a night attack, stops talking, and disappears outside. The relevant captain takes down every word he says in a notebook and sees to it that orders are carried out. At the beginning of the briefing, DePuy introduces me to his officers in a very complimentary way, asks if I have any questions, and I say no.—Everyone hurries to prepare. I am also shown the cistern (into which I should jump and use as a bomb shelter in case the camp is attacked). I am given a flashlight and helmet; they explain the signs to me—a siren, etc.—and the password.

A South Vietnamese officer riding on a Lambretta with four other officers in the vicinity of the camp was taken captive by the Viet Cong. They held him captive for several days, until a division unit landed near them yesterday. He took the opportunity to escape during the commotion and is now eating greedily (they did not feed him in captivity), receiving medical care, and handing over details.

Amongst other things in the briefing—another bridge has been blown up. Five soldiers riding in a jeep have been killed by a mine.

During my conversations with DePuy, I tell him about our past situation in Gaza and my conversation with the mayor about work stoppages, etc.—If this would occur, their own people would be hurt, not his own son. I see that my words impress him (He tells me that a way must be found to prevent war with Americans from hurting Viet Cong civilians).

During the briefing, I tell the publicity officer that announcements should be sent out, explaining that, when the Viet Cong do things like demolishing roads, they don't hurt Americans (who travel by helicopter); the only ones who suffer, are Vietnamese civilians.

The Viet Cong move around in the jungle in small groups (DePuy differentiates between a jungle of trees and bamboo, which is much more difficult to move around in: he doesn't call bamboo a jungle). They know the area well, fix a meeting place, and then travel there in small groups on scattered roads.

In today's newspaper, Johnson announces that, if Ho Chi Minh stops sending soldiers to the South, they will stop the bombing. This does not correspond to what General Larsen said.

I share my opinion with DePuy that, within a year or two, both sides will reach a kind of 'modus vivendi'—the large and small potatoes in the sack will toss to and fro until the large push the small potatoes aside; the small ones with settle as best they can in the nooks and crannies. The Viet Cong will not be able to stand the competition, but they will find cracks. They will hide in the jungle, obtain arms, etc. from abroad, and choose easier guerilla targets. They will fight with the South Vietnamese government as if they were Quislings,[43] and find ways other to prove to the Americans that their opposition will continue. The Americans want to conclude the war—by normalization. But, against this fact, the Viet Cong can forget about getting rid of the Americans (as they did the French), but prove that they are still fighting, and not make peace.

Washington will also need a new conceptualization of the war. McNamara and Taylor[44] told me that:

1. More than 1,000 Viet Cong are being killed every week.
2. North Vietnam will not be able to withstand the bombings.
3. The (pacified) area under American control is getting bigger. But:
   a. It is likely that the North will make peace with the situation (and return to primitivism, despite the theory that socialism is proud of its technical achievements and cannot relinquish them).
   b. The Viet Cong will stop its futile attacks, and the number of killed will diminish.
   c. 'Control' over people and areas has no significance whatsoever. Therefore, Washington's assumptions are not based on fact.
   d. The real 'tragedy', is that life is not ordered, with male and female animals peacefully entering Noah's ark two by two as described in the Bible. In real life (wars), some problems are insoluble.

In today's newspaper—the Viet Cong have evacuated all those living in a refugee camp and set fire to it. They have no lack of such targets. Such an act will remain legitimized in the eyes of the public, as long as the Viet Cong can point out the activities of the government and the American as motives for the war.

It is midnight. Perhaps, despite the shooting and the heat, I will still be able to fall asleep. In the meanwhile, the light in my tent has attracted thousands of insects. Azaria Alon[45] would have a field day doing his entomological research here.

22 August 1966
Lai Khe

During the night, I anoint myself with a double layer of mosquito repellant. The problem of sleep is twofold: mosquitoes and cannons. Inside the camp, the ground and walls shudder with each impact (mainly of the 175mm guns). Apart from that, one must be alive to the difference between 'outgoing' shells—that are being fired outside the camp—and 'incoming'—Viet Cong shells that explode inside the camp.

DePuy guesses wrongly—his camp is not attacked by Viet Cong infantry. The morning program is therefore to visit the division's 'pacification' activities.

But, before that, something else comes up. The patrol that went out yesterday has lost radio contact, and division headquarters have heard nothing from it. When DePuy comes in for breakfast, after a quick 'good morning,' he calls the two senior colonels to his table, and asks them whether they knew that the patrol (eighteen men) has not returned: he has not been informed that they have lost radio contact (his words are accompanied by a vigorous thump on the table with his first). 'This is the most serious thing that has happened to this division since I took command.'

During breakfast, it becomes clear that the patrol has returned safely that morning. But DePuy refuses to let the matter rest at that. After reprimanding his officers for not being more attentive to the fact that communication with the patrol had been cut off, he goes out to the division rear headquarters—from where the patrol had been sent out. Rear headquarters (20 minutes by helicopter) consists of wooden, stone and tin buildings, scrubbed clean and well organized. The office of the division commander (DePuy) really sparkles. (He explains that he has never been there before). According to his command, all soldiers at rear headquarters must—without exception, officers and men alike—participate in a military operation every two weeks—patrol, guard duty, weapons exercises, etc. In the beginning, the men grouse about it—letters of complaint to the 'inspector-general' of the army—but after that they concede.

In my presence, the lieutenant in charge of the patrol, and all his superior officers, are called into DePuy's office. It turns out that:

1. Two pieces of communications equipment had stopped working (DePuy fixes a steely-eyed glare on the officer: 'I don't believe it; if the batteries would have been changed, the equipment examined before the patrol left, and the radio operators not been asleep—two perfectly good pieces of equipment would not have broken down at the same time: this teaches you to do a proper prior inspection').
2. Despite the communications problem, the patrol commander decided to remain until his assignment was complete, and return in the morning.
3. When communications were severed, they informed DePuy's deputy, General Deane,[46] who was here at the time, and not at division headquarters. DePuy examines standard operating practice in such cases—there is none.

This must be remedied at once. DePuy instructs: In such a case:

1. The patrol should return immediately (he says quietly to the younger officer: 'without radio communications, you could not obtain artillery assistance').
2. If headquarters sees that there is no radio contact—they must send out immediate assistance. If they knew about this last night, they should have sent a battalion out to look for the patrol.
3. Lighting methods (flares) must be put in place, as secondary communication tools.

The most difficult part of his words is his in-your-face reprimand to all groups of officers in the field—'I see this as an extremely gross error. It is unprofessional in the extreme.'

I try to soften his wrath a little, saying that, in any event, I respect an officer who decides to remain in a place without radio contact over one who immediately loses his cool. If he learns, this young man will make a good battle officer.

DePuy really respects Viet Cong soldiers. He tells me of an incident when ten well-dug in Viet Cong delayed one of his battalions. After a prolonged battle, nine of them were killed, and only one was left in a bunker without a firing position. Each time they dropped a grenade inside—he threw it back. If he couldn't do that, he covered himself with a body in the bunker. They eventually killed him by first freeing the detonation handle and then throwing it in after a few seconds delay—it exploded immediately.

He tells me that the Viet Cong sometimes open fire at helicopters passing overhead, even though by doing this they attract artillery and air fire. They do this even though they are in small groups and know that they are about to be killed.

We fly to the region of Dong Nai—20,000 inhabitants (a town with several adjacent villages). The area was previously under Viet Cong control. According to DePuy, they have an organized staff (50% as big as the current regime and 200% more efficient), with an agricultural representative, village chief, etc. I ask whether their representatives only take. He says—no, they also give. Everything that they can obtain from the central (Saigon) government—including artificial fertilizer donated by the Americans—they hand over to the farmers. Their rule is not corrupt. Fifty-two of fifty-eight nearby villages are under Viet Cong control.

The aim of 'pacification' is not to kill the Viet Cong, but to convince them to become government supporters, by changing cadres—leadership—and proving the independence of the government. It is hoped that they will follow the stronger side—or know that they are not here only temporarily.

The man responsible for 'pacification' activities in the division is a battalion artillery commander (who is leaving in a weeks' time). He explains that they have combed the area out thoroughly, and imprisoned Viet Cong, deserters etc. They are now patrolling the town with five American units—five to seven men in each patrol. They are in radio communication with reinforcement units. The schools are functioning, and there is medical aid.

We visit a school—a principal, and several female teachers (DePuy is surprised at this, but I am not)—they are the same teachers who taught here when the school was under Viet Cong supervision (probably the same principal as well). There is now a new village chief, living in the village (this is an important point, because village chiefs are a central target for Viet Cong terror), under the protection of American soldiers. His family (this surprises DePuy as well) quite obviously does not live here.

There is a long line of women, (old) men, and mainly children, at the clinic. The children suffer from all sorts of scrapes, wounds, blisters, etc.—mostly from lack of hygiene and jungle conditions (insects, skin rashes, etc. Heat and humidity also play a role). Those providing treatment—are both Americans and locals. There is no doubt of the great importance of this activity—from a humanitarian point of view.

The battalion commander (responsible for this operation) says that—if the battalion would leave the area—the situation would revert to its former state. The battalion covers an area of 10 x 20 kilometers. Two divisions are necessary to cover the entire province (Vietnam consists of forty-one provinces): one American, one South Vietnamese. Their division is about to add one more battalion to this area—for guerilla warfare. The coming period is of special importance: the elections. The Viet Cong have announced that anyone who participates in the elections—will be killed. The elections will therefore be a test for the government.

'Cadres' called 'black shirts' walk around the town, dressed in black shirts and carrying arms. These men have graduated special courses designed for cadres (leadership) and are sent to the towns and villages to serve as a nucleus for a pro-government administration. Of course, they live courtesy of the Americans, who protect them.

Meanwhile, DePuy is handed a report that heavy fire had been directed at his helicopter. His deputy, General Deane, is flying in it, to spy out the road at low elevations ('Route 13')— looking for local Viet Cong tax-collecting barriers. The soldier operating the machine gun is killed. DePuy says that pilots must fly higher, above tree level, to make it difficult to hit the helicopters. We return to camp. We pass by the damaged helicopter at the airfield: its floor and walls are stained with blood, and there is a red pool of rain and blood under it.

General Westmoreland
22 August 1966

At 2.30 pm, Meeting with Brigadier General Jones, General Westmoreland's chief of staff.[47] A kind of feel-each-other out conversation—a question-and-answer session. His answers to my questions:

1. They believe that the coming stage—opening the roads to traffic—is a precondition for normalization, and that this can be done with the aid of additional forces he has received.

2. The army does not have a map defining the three areas—under their control, under Viet Cong control, and mixed. But he thinks that Welfare Administration [USAID] has such a map. He himself agrees that it is impossible to exactly define who controls what. While the Americans are in the area—the people are susceptible to their influence. After they leave—it will return to the Viet Cong.

3. It is correct to consider those who live in a Viet Cong region as hostile—in other words, it is allowed to fight there. He realizes the defect in his reasoning—but war is war.
4. His forces (together with those of the South) will be enough to achieve the goal of pacification. They have now reached the stage that guerilla warfare is their first preference.

In the evening (8.00 pm)—dinner with Westmoreland (I give him a copy of my book on Sinai). Two other generals are present—the guest from the Pentagon (whom I met with General DePuy), and a young man who has returned from leave and is apparently on Westmoreland's staff. All of us are dressed in open-necked, short-sleeved shirts. Westmoreland is a pleasant but unimpressive man with an affable countenance.—Neither stubborn, nor as intellectually shrewd as DePuy and Larsen. Of course, it is not possible to draw proper conclusions from a 2-hour conversation.

Immediately, with the first drink, he asks me straightforwardly about my impressions and conclusions of what I have seen here. He tells me that he had received a request from McNamara, and an order to all his staff:

1. To facilitate my visit;
2. To look after my safety;
3. Not to allow me to go out on patrols.

First, I thank him for how all his staff have treated me: their courtesy and cooperation are more than I deserve. He is happy to hear it, and I understand that he wants McNamara to hear this as well. The meal and discussion last for 2 hours, and I request to leave. Before I go, we arrange (at his request) that I should return and see him again one more time before I leave Vietnam. I intend to simply say goodbye, but it seems to me that he wants to talk more.[48]

I cannot help adding my main conclusion points. It is possible that I voice my criticisms too roughly—especially when our opinions differ. My main issues are:

1. If the Viet Cong and North Vietnamese forces continue with their policy of military clashes, with the purpose of defeating the Americans, as they did with the French—they (the Viet Cong) are lost. The Americans are very strong, and the Viet Cong have no possibility of beating them by classic military means.
2. The American Army are not only not tired of this war (not all of them—and of course not the privates) but are enjoying each day of fighting here. (When I say this, the general urges me: 'write this down': I say that I will, but that not everybody in the United States will be as happy to hear this as he is. Intellectual circles would find therein a reason to defame the reputation of anyone who enjoys making war).
3. The Viet Cong may conclude that they must change their tactics and go over to guerilla warfare (both here, and during the continued discussion, Westmoreland is of the apparent opinion that they would continue as before and try for a straightforward military victory over the Americans).

This gives rise to other questions. Westmoreland: 'An army can defeat guerillas, but guerillas cannot achieve a military victory.'—I want to tell him that this is only a 'cliché,' but restrain myself and clothe my words more elegantly: guerillas do not have a uniform fighting template: In this environment, they have the following advantages:

1. They can hide in the jungle. The jungle can be an excellent haven—despite all American resources.
2. They can have withdrawal bases: Cambodia, Laos, North Vietnam.

3. They may have outside help (money, ammunition, foodstuffs, etc.) from China, the Soviet Union, and perhaps other countries.

Secondly—guerillas cannot 'beat' the army, but the American war objective is peace, and normalization of everyday life. By contrast, guerillas can be satisfied with demonstration of opposition, non-acceptance, and continued war status. For this purpose, they can select convenient targets, provided that the (their own) public accepts them: roads, 'Quislings,' ('black shirts'), 'hit and run' ambushes. attacks on army units, etc;

Westmoreland explains that the American war here is like a boxing tournament. With one hand, he must forcefully strike (the regular army). But at the same time, he must protect himself with his other hand from attack by guerillas, while he himself is open and exposed. He might be fighting the army, when guerillas strike.

My criticism points:

1. Some of their patrols (with big units, larger than platoon-size) are badly planned. They are too heavy to reveal the enemy, and vulnerable to ambush. This even happens to the Australians, professional jungle fighters (Westmoreland agrees with this completely).
2. They are sometimes too aggressive, instead of waiting and maneuvering, etc., they move about a lot searching for the Viet Cong—and sometimes the cost is more than it is worth. The problem here is not to 'walk kilometers,' but to do real damage. Aggressiveness without carefulness—causes failure.

Westmoreland: 'In war one must take risks' (another cliché).

Dayan: 'When we were fighting in Sinai, we were forced to take risks. Had we not finished after 5 days—you would have stopped us. Risks had to be taken with the war against the Germans and Russians as well. But here, you have no limitations in time, area, or maneuver. You can plan and operate at your own convenience. In this regard, I assess positively the enormous firepower that you bring to bear on behalf of each soldier, by bombing Viet Cong snipers with B-52s.' (I mention that France had lost an entire generation of youth during the First World War, and quoted Montgomery about the loss of a million young British men during the Second World War.[49] Bombings avoid this). 'But the same logic obliges you to prevent excessive losses resulting from lack of proper consideration, and too much aggressiveness.'

Carpenter[50]—who requested that his men 'fall at their positions'—has been praised and honored. I consider what he did to be—a crime. He should have been court martialed, not decorated. He had wounded under his command. How could he ensure that his own men would not be killed by friendly fire? (Westmoreland does not react but appears to agree with me).

Even the Australians, whose last operation he praised so highly, would have been in a bad way if not for the arrival of an armed convoy, almost by accident.

When Westmoreland praises the Korean unit for their recent success—I say that, with all due respects, the Viet Cong were really subdued by 21,000 shells, bombings from the air, and tank companies—all of them American.

The main bone of contention this evening is the issue of 'pacification.' I am not sure that I succeed in clarifying my stance (although, in my own eyes, it is very clear).

I say that, in my opinion, their plan of 'pacification' is exaggerating the 'Americanization' of civilian life here, thereby creating convenient Viet Cong guerilla targets. I think that they should decrease their activity in civilian areas as much as possible. Assistance should be given indirectly. A situation must be created whereby, if the Viet Cong sabotage the roads—it will be regarded as a war against their own people, not the American foreigners.

I give three illustrative examples:

1. The (female) teachers. I am happy to see a 'neutral' school. If the teachers (who served in the days of the Viet Cong) would have

been exchanged by American teachers—the school would have been turned into a guerilla war target. It is better to leave the teachers where they are, removing the school from the area of conflict. This, even if what the teachers taught favors the Viet Cong.

2. At the gate of some northern villages, I see two forlorn sentries from the civilian population, rifle in hand, 'guarding' the village. The Viet Cong can obviously overcome and kill them, but this would create a clash with the village, for the purpose of stealing rice from it. These guards must be given the possibility of communication (only this), to be able to summon American help. The struggle must be turned, as much as possible, into one between the peoples of Vietnam—government and Viet Cong. The American patrols in DePuy's small town are not a solution to this, but will rather complicate matters, changing a civilian issue—ordinary daily life in the town—to one of a fight against foreigners, Americans. I think it better not to enter this area, but, again, only as a 'response,' coming to help.
3. In this spirit, if they could (I don't know if it is possible at this moment) turn the question of Vietnamese roads into a civilian issue—in other words, road damage doesn't hurt the Americans (who can do all their travelling by air), but the population itself—this would be a step in the right direction.

Another example: They must leave in place Viet Cong clerks who are prepared to continue working even if the region would come under American control. It is better not to fire someone appointed by the Viet Cong for agricultural assistance—if he wants to continue under a new village head—even if he may not be loyal to the government. The area must be left as an internal struggle between different Vietnamese groups. The Americans must decrease their sphere of activity, and position it as a war against military units, and the forces of North Vietnam.

Another example—not for comparison, just to explain my thoughts—I cite our situation when we conquered the Gaza Strip. The Arabs immediately proclaimed a strike of schools, stores, etc. I called in the mayor and told him: Our army will remain outside the city. If you close the stores—your own people will be the only ones to suffer. The same applies to closed schools; this will harm your own children. We will not interfere. As conquerors, our duty—which we will fulfil—is to help you. If you want sugar, flour, etc. from us—it is our duty to give you sufficient food, which we will do. If you need our army's help against thugs—we will provide it. If you need financial assistance for your school, hospital, etc.—we will provide it. But if you wish to close these institutions—we will not open them by force, nor will we turn civilian institutions into a struggle against foreign conquerors.

I stress that I know the situation here to be completely different—here, the Viet Cong are fighting against the government of Vietnam. But despite this, ways must be found to diminish the Americanization of civilian life and allow a compromise to be worked out between the local Viet Cong and the government, based on recognition of American presence here.

I say that, as far as I knew, most of the 'old guard'—the allies who fought along with the United States during the Second World War—do not want America to fail, with a resultant Northern victory. This, despite all their criticism of the way the Americans are prosecuting this war. The Americans are strong and have advanced so far, compared to their allies, that there are now left with no allies at all—they are alone.

I doubt assumptions by Taylor and McNamara that Ho Chi Minh would be forced to the bargaining table. It is likely that North Vietnam will adapt to a primitive lifestyle—without electricity, roads, centers, etc.—but still not submit. They will diminish the number of soldiers here and prevent heavy losses (this week alone—1,500 killed).

23 August 1966
Saigon

`I have dinner with a Jew named Grushka. He was been in Israel for 2 years; his brother is with Baram (Mapam).[51] His wife is Vietnamese.

The French vice-consul in Saigon joins us for dinner.[52] He is also married to a Vietnamese woman and has family in Hanoi. He tells me that he receives letters from Hanoi—via Cambodia. His main points about the situation in Hanoi:

1. The people receive food with ration cards. These cards are only for workers, Communist party members, soldiers, etc. Someone with a ration card receives 140 gm rice daily, and that is it. No meat, no fruit, no fish—nothing.
2. (Most of) the population of Hanoi has been evacuated.
3. They live in permanent fear that the Americans will bomb the dike—30 m[iles] above Hanoi.
4. The Chinese are not helping North Vietnam, but the Russians are. Relations between Hanoi and the Chinese are bad. The Chinese do not value the brand of communism in Hanoi, and there is an age-old enmity between the two countries.

Conditions here: De Gaulle's hatred of the Americans is pathological.[53] They pray for an American victory, because woe betide if the Chinese and Vietnamese Communists win. Added to this—the Americans committed a fatal error (Cabot Lodge) when they engineered the murder of Diem. There is no one to replace him, and they will not be able to establish a real government here.[54]

Endnotes

1 This is likely a printing error in Hebrew text. Đức Phổ is a district town in South Central Vietnam known as Đức Phổ Base Camp (also Đức Phổ Airfield, LZ Bronco, LZ Montezuma). Đức Khổ was a Vietnamese historical figure known for his role in the anti-colonial struggles against French rule. Đức Khổ was a prominent leader in the early twentieth century who played a significant role in Vietnamese nationalist movements. His contributions were part of a larger context of resistance against colonial powers in Vietnam, and his efforts are remembered as part of the broader struggle for Vietnamese independence and sovereignty

2 The Civilian Irregular Defense Group (CIDG) was created by the CIA in 1961 to help develop units of South Vietnamese Special Forces.

3 Stanley R. Larsen (1915—2000) was a United States general who served as commander of I Field Force Vietnam during the Vietnam War.

4 Galilee Area where swamps were drained.

5 This is a contingent from the 16th Armor. Captain Wells Eldon Cunningham was killed in action 17 August 1966; Mountain people: Montagnards (French).

6 Robin Moore, *The Green Berets* (New York: Avon Books, 1966).

7 This unit may be the South Vietnamese 3rd Armored Squadron.

8 Francis J. Kelly (1919–1997). He was born in the Bronx and first enlisted in the Army in 1941. He received a commission in the Armored Corps in1942 and was among the first wave to go ashore on Omaha Beach on D-Day (June 6, 1944) at Normandy. Among his many decorations were the Silver Star and Bronze Star. He was a graduate of the University of Maryland and Georgetown University and after retirement, earned a doctorate from the University of Denver.

9 David Daniel (Mickey) Marcus (1901–1948), United States Army colonel who played a significant role in the Israeli War of Independence, building a road to Jerusalem that relieved the siege.

10 Chapter 1, notes 21,32.

11 Chapter 1, note 50.

12 Translator's comment: many of the points raised are opinions, not facts.

13 Translator's comment: General Charles de Gaulle was a complex personality, not easily understood even by Frenchmen. He was certainly not inherently Anti-American.

14 Chapter 1, notes 29, 48, 67.

15 Tuy Hòa; Dayan's reference to the 10th Brigade, is incorrect. It should be 1st Brigade, 101st Airborne.

16 Lieutenant General Willard Pearson, USA (1915–1996).

17 See note 1.

18 Chapter 2, note 2.

19 Jonathan Swift, *Gulliver's Travels.*

20 He is implying that Viet Cong loyalty shifted occasionally, and Viet Cong troops joined the South Vietnamese Special Forces at camps such as Duc Co, to be trained by the CIDG. Dayan may be referring to the efforts of the 'Chieu Hoi Program' which was established by the United States and South Vietnamese governments to encourage Viet Cong soldiers to change their allegiance and support South Vietnam. Whether 'whole units' ever engaged in such activity is unlikely.

21 Translator's comment: A statement of profound importance.

22 See Chapter 3, note 14.

23 Named after the Maginot Line, the failed defensive line of border forts built by the French before World War II.

24 This would have been an Army medevac helicopter which was marked with a red cross.

25 This is likely a reference to the M79 that fires a 40 x 46 mm grenade and was first used in Vietnam by the Army in 1965.

26 Including Agent Orange.

27 Võ Nguyên Giáp (1911–2013), was a general of the People's Army of Vietnam, communist revolutionary and politician. Regarded as one of the greatest military strategists of the twentieth century, Giáp led the Viet Minh Army during their successful attack at Dien Bien Phu. A dedicated reader of history and philosophy, he revered Sun Tzu and successfully used *The Art of War* against the American military.

28 Chapter 1, note 47.

29 *Notrim*. Jewish Supernumerary Police.

30 Barry Zorthian (1920–2010), was an American diplomat and chief spokesman for the U.S. government through the U.S. Information Agency in Vietnam from 1964–1968.

31 Willam J. Porter (1914–1988), was Deputy Ambassador to South Vietnam (1965–1967). He had previously served as Ambassador to Algeria, and in 1971 he became head of the U.S. delegation at the Paris Peace Talks.

32 Dayan is referring to the United States Agency for International Aid (USAID).

33 Translator note: Quite obviously nonsense.

34 A conservative British daily newspaper.

35 Chapter 1, note 48

36 Bertrand Russell (1872–1970), British philosopher, logician, mathematician, historian, writer, social critic, political activist and Nobel laureate.

37 Translator's note: A very condescending remark.

38 Chapter 2, note 14.

39 William E. DePuy (1919–1992). First deployed to Vietnam in 1964, DePuy served as Chief of Staff of Operations for Military Assistance Command, Vietnam, and in March 1966 he assumed command of the 1st Infantry Division.

40 Lai Khê.

41 In 1966, Nixon was positioning himself for another presidential run in 1968.

42 Australia participated in the Vietnam War as part of a United States-led intervention to Vietnam to assist South Vietnam against North Vietnam

43 Chapter 3, note 37.

44 Chapter 1, notes 10, 57.

45 Azaria Allon (1918–2014), noted Israeli environmentalist.

46 John R. Deane, Jr. (1919–2013).

47 General Westmoreland's chief of staff for much of his time in Vietnam was General Frederick C. Weyand.

48 Translator's comment: Westmoreland obviously wants to continue discussions with Dayan.

49 This is what the original says. Obviously, the number refers to World War 1. British losses in World War II were approximately 1/3 as high.

50 Captain William S. Carpenter, Jr. (b. 1937, retired a Lieutenant General). During the Vietnam war, he was awarded the Silver Star which was later upgraded to the Distinguished Service Cross for bravery. Dayan's remark appears condescending.

51 *Baram* is a kibbutz program, and *Mapam* is the name of one of the old Israeli labor parties.

52 See Chapter 1. France did have a consulate in Saigon during the Vietnam War, although diplomatic relations between France and South Vietnam were severed in 1965, meaning the consulate essentially functioned as an embassy in all but name during the later years of the conflict

53 Untrue. Charles de Gaulle was a very complicated person.

54 President Kennedy was assassinated on 22 November 1963 a few weeks after Diệm's murder on 3 November 1963, leading to a series of unstable military governments.

— 6 —

DEVELOPMENTAL AND AID ENTERPRISES

United States Agency for International Development 24 August 1966 Can Tho

This morning, I begin the civilian portion of my visit. Although I have only 2 ½ days left, I believe I'll be able to see the main issues of importance to me: refugee settlement, care in existing villages, and special activities—physical or organizational—irrigation, farmers loans, organized marketing, etc.

The United States Agency for International Development controls all civilian activities—clothed in different forms and names. (Americans love abbreviations, catchwords and sensational headlines). If I understand correctly, activities are divided as follows:

1. 'Civilian or humanitarian activities'—Medical assistance, schools, care of refugee camps. All these are carried out by army units, United States Agency for International Development, and local administration.
2. 'Pacification'—control of civilian security—war against guerillas, direction of institutions loyal to the government, police, etc.). This is done by American and Vietnamese armies, and local government (preparation of the cadres for leadership and administrative

roles). The Agency for International Development comes into the picture after the area has been more or less 'pacified.' However, army units also perform civilian work during their military duties 'for pacification purposes.'

3. 'Revolutionary development'—A new name created in Honolulu at a meeting between President Johnson and the advisors of the Prime Minister General Ky.[1] Before this, they called these activities 'village development.' Their meaning is—promise of 'a new way of life'—agricultural and economic development, etc., to raise the standard of living. This is done by the Agency for International Development and local government.

The Agency for International Development is staffed by professionals, who work on contract for shorter or longer (18 months) time periods. They 'implement American foreign policy' (but are not idealists, full of all sorts of ideas and concepts). I assume that they are paid well: their standard of living here is satisfactory, or least it seems so to me from their apartments, with dining room, etc. At dinner I am the guest of the man responsible for this region, Colonel (I think on leave from the army) Harry Jackson. (By the way Westmoreland's[2] Saigon lodgings are—pleasant). A large house, garden, furniture, etc., in colonial style (built previously by the French, for their high-up officials). The only thing that it seems they cannot get rid of is—insects: small flies and beetles that crawl into every apartment on every floor. The ground, the air—are full of them. I am also happy to see several lizards crawling around the walls, sometimes hunting one of the small insects with their long sticky tongues.

At 8.30 am I receive a briefing in the building of the Saigon Welfare Administration [*sic*],[3] from Nyugen Ro Thanh I believe that he is the deputy—or one of the deputies of—the director, who is currently out of the country). I don't learn much that is new. Their main job is to deal with ongoing economic issues, the greatest of which is inflation. The war against inflation is fought by imports from the United States: the piasters with which local merchants pay for imported goods go to the government budget—about 500 million dollars in the current year.

Aid experts are usually integrated into different government ministries, as 'advisors.' With their financial influence and status, they direct all governmental activities (except perhaps for religious-ethnic policy, parties, etc. In any event, they direct all economic activity). Those in charge of agriculture also participate in the briefing. There are currently about 2 million families working in agriculture, in other words 10 million souls, corresponding to 2/3 of the population. By contrast, there are about sixty Americans dealing with agricultural issues, in four main spheres of activity:

1. Introduction of artificial fertilizer for day-to-day use. They believe that half a million tons are needed each year, but this year only 250,000 tons will be provided (50% of the country's needs). Part of this is given out free, but most is sold to the farmers at half-price.

2. There is elimination of insects by spraying.

3. Research—varieties, seeds, cultivation methods, etc.—There are seven research and experimental stations.

4. Species improvement—rice, vegetables, fruit, pigs, cattle, etc.

The issues of refugee settlement, and their exact number, remain unclear. Initially they numbered about a million who fled from North Vietnam (Hanoi) to the South. Most of these refugees are now in the cities, where they somehow eke out a living with and without governmental aid. The war has now added another million: about half of these are 'settlers'—in other words have been absorbed into all kinds of work. But the main issue is that they represent a decreased government burden: they are self-sufficient and help each other.

At this time, there are still about half a million refugees on governmental assistance. Their number increases 20,000 to 30,000 every month (it is not clear whether this will continue).

My plan includes a visit to a village where refugees were settled in agriculture. I ask that my program be changed to include two or three more villages (instead of a visit to a hospital, school, etc.).

The main disappointment at today's briefings is the lack of long-term planning. The Welfare Administration works on an annual planned budget, with a general 2–3 year strategy. Clearly, the plan is how to forecast future problems and suggest their solutions, not an economic design to change the standard of living, market structure, etc.

In favor of the Agency for International Development, it must be said that the country is in a state of war, and that peacetime economic planning cannot be implemented. But, despite this truth, I am convinced that they should have a long-term plan for at least a few issues and try to act accordingly.

Two other things do not work out during today's inquiries. The first—is a map of regional control. In Washington (also according to General Taylor's[4] testimony to Congress and certain groups here), percentages are used to denote regions under control of the government, the Viet Cong, and in-between. I request a map showing these divisions and am promised one for the Delta region—which is the most stable. However, I am sure that all this division into percentages and attempts to fix the rate of war progress based on increasing percentages of government control—are quite meaningless. There is no doubt that the government's area of control has increased, compared to what it was last spring, but it is still not a matter of a map and percentages. The American Army controls where it is physically present. However, if it leaves the area, control simply returns to the Viet Cong.

The 'exact' numbers on Agency for International Development's achievement in common country-wide use of artificial fertilizer are not based on fact. I can only attempt to clarify numbers in the Delta region. This year's crop has diminished—by about 5%. Part of this decrease may be caused by neglect of entire areas (? 100,000 dunam[5]) in the war zone (this is not at all clear, though). By contrast, about ¼ of the area receives artificial fertilizer which should, theoretically, increase crop yield by 30%. But, instead of increasing—it has decreased.

I believe that the reason for this lies in statistics. The farmers sell part of their crops on the black market. The merchants do not declare their full inventory to the government (official government prices are lower than those on the black market). Therefore, even if crop yield has increased, this is not properly reflected in the government's books or in statistics.

In real life, artificial fertilizer does not always play a role. Water washes it away, and there is a certain amount of growth (the tall rice planted in water) which is not influenced by fertilizer. I think that McNamara[6] will be somewhat disappointed when he hears the truth about his chemical fertilizer. The most prominent merchants here are Chinese. They buy the rice from the farmers, paying them in advance at the beginning of the year at 25% interest—that is the official interest rate here!

A Quiet Area
24 August 1966

I fly from Saigon to Can Tho in a Piper leased by the Agency for International Development Welfare Administration. We fly at high altitude: the pilot explains that clouds prevent him from flying at intermediate altitude; below the clouds, we would be fired upon by the Viet Cong. The entire area is a crossword of squares and murky rivers, and endless canals which cut through them in all directions. Some of the 'squares' are green, others ploughed, waiting for the next planting. Yet others are flooded with water, like large square lakes. Fishponds, in which carp are raised, are common here.

When I arrive, Jackson begins his briefing. This is the quietest region in Vietnam. Both the town and the main road are full of civilian traffic (during the day: at night, there is a midnight to 5.00 am curfew). Sometimes there are mine explosions, etc.—but the traffic continues as usual for more than 100 kilometers—from the Cambodian border to here (I don't know whether the same is true from here to Saigon). Things are relatively quiet (however, if they leave the main road—security ceases, and they reach areas under Viet Cong control). There are different reasons for this road safety. The main one is—the government has succeeded in overcoming (killing, etc.) the Viet Cong. Also, the terrain (a plain without jungle) makes it difficult for the Viet Cong to hide. My guess is that there is yet another reason: simply—refusal to be beaten, refusal to give up use of the roads. The population carry on using the roads under all conditions. This has created a situation where roads have been removed from the field of battle. The road here is the farmer's staff of life, and it is not easy for the Viet Cong to harm them too much. I have lunch with the local Agency for International Development staff, also with those in charge of the Vietnamese army (and their American advisors). Senior Vietnamese officers are also present.

There is no doubt that local (Vietnamese) rule is stronger here than in other places. Their purpose—is to replace Viet Cong administration with their own men. There is no lack of men who want to obtain government positions, and do not want a compromise with the Viet Cong. They feel strong enough to defeat the Viet Cong and remove them entirely from their areas of influence (they are not prepared to leave female teachers in place, like DePuy[7]). All this is crucial to the fact that this whole area is under complete Vietnamese Army control, without Americans (In the same overall sector that is under government control, however, significant areas in the Delta are controlled by the Viet Cong). About half a million civilians live there, controlled by:

1. A 400-man strong Vietnamese battalion, with its own artillery;
2. 300,000 regional forces. They, in the fullest sense of the word, are an 'army of the people';[8]
3. 3,000 local forces;
4. 700 police.

There is a total of about 7,000 Vietnamese soldiers, with only twenty-five American advisors.

The Vietnamese forces are of course able, at any moment, to receive American air assistance, and—if necessary—reinforcements. But, as things now stand, this region is ruled by local Vietnamese forces!!

In contrast to what DePuy said, they maintain that the Viet Cong organization is wretched and valueless. They conquer an area—but there are no proper schools, etc.

During the afternoon, we visit some farmers. A small island in the middle of the river. The area is saturated with water, and embankments, etc. need to be erected to preserve the land and allow it to dry out. The land is (apparently) owned by the government, who lease it to the farmers. A 'land unit' is about 4 dunams.[9] I visit an exemplary agricultural settlement (a kind of government model farm) for growth of vegetables. They now have beans, radishes and rice. The instructor is an expert, who works for the government. On one side, everything looks very primitive, and the standard of living is extremely low (the 'house,' 'beds,' and total means of making a living are 4 dunams). But, on the other hand, it is impossible not to be impressed by the fastidious agricultural and garden cultivation—each plant receives 'personal' attention, and the crop yield must be high. I am not surprised—but such agriculture will not bring them redemption, neither will it raise their standard of living.

Amongst other things, they are building a system of pipes on the island, to regulate the floodwaters and the river. Pipes are poured from concrete, with an approximately ½ meter diameter and an iron foundation. But, instead of normal structure in the inside portion, they use split layers of round, 'natural' banana tree trunks.

In our honor, we are given bananas, and fresh coconuts in which holes have been made. The juice is tasty and cold. Our American escorts are extremely impressed, and there is no doubt that the picture is impressive. We reach the island in a motorboat, through dense growth of jungle rubber plants. The farmers—legs swimming in mud—are very agile with their hands, and their diligence is impressive. But all this reminds me how impressed the British were with the Bedouin way of life—instead of improving them and introducing them to a more suitable standard of living.

Two squads of Vietnamese soldiers accompany us, to make security doubly sure. I am not impressed with their 'exemplary' order: their leader doesn't stop yelling and berating them.

We travel further, to visit an agricultural middle school, where the pupils are taught at two levels: the first lasts 2 years, and the second 3 years. It takes 5 years to reach the upper level.

We visit their school 'farm.' Some miserable fruit, a tiny chicken coop packed with all kinds of multicolored hens and roosters, different sorts of technical implements from 50 years ago, two tractors, a few vegetables, a few fruit trees, etc. It is really a disgrace. The school has 600 pupils! And after this they become 'agricultural experts,' absorbed into the governmental organization.

In summary, I am very disappointed with the first day farm tour. Of course, the country is at war: But I don't see a single sign of planning, modernization, or correct agricultural thinking. Even in wartime, long-term planning is necessary. The school should have a modern chicken coop, cow barn, and demonstration plots. There is no movement in that direction. The water buffalos are very impressive, with their muddy color and enormous horns—but to rear them as plow animals in an 'exemplary' modern agricultural school in the Delta, a school founded by the French about 40 years ago, and now refurbished by the Americans—where is all that culture from Versailles and California?

It should be possible to establish a large government farm here, to grow rice and vegetables. The military will have to secure the area, use modern growth methods—under American supervision, and so direct that, in the fullness of time, it becomes a village, handed over to its workers. At first, the workers will be hired by the government—they are not short of refugees, land, water, a market for vegetables, rice, meat, fruit and eggs. A few American and local experts can do this. Such a farm, with a school, clinic, and other village institutions, a cooperative marketing store, can certainly be self-sufficient, bring enough rice and other products to market, and be cultivated with modern methods—correct artificial and other fertilizer and spraying. Land units are twice or three times larger here than is usually the case. And something else must be taken into consideration—now, most of the young men are in the army (on one side or the other), and this is an opportunity to create a large amount of produce with fewer working hands, get rid of hidden unemployment, and begin proper mechanization. When, God willing, the young men return from the army, they must be able to be absorbed—in large land units (increased land by dint of jungle training), industry, jobs in the city, etc. The time is now, and it is possible to increase production.

The entire staff—several dozen people—assemble for dinner—doctors, nurses, all kinds of public workers. In the background—music and dancing. A dimly-lit bar, 'jungly' branch roof, martinis, whisky, etc.—a 'superior' buffet, everything courtesy of the Agency for International Development staff.—Vietnamese only as bartenders. A kind of colonial administration styled on America and its aims, with the obvious stamp of colonial administrators working to achieve the goals of the mother country—self-satisfied, with a feeling of usefulness, rule and mission. I myself do not feel as comfortable. This is the first night during which I go to sleep without the accompaniment of cannon fire. Even in Saigon, if one listens carefully, echoes of firing can be heard. Here complete peace and quiet reigns.

The Delta
25 August 1966
Can Tho

In the morning, I am taken to (Vietnamese) army headquarters, to look at (but not receive) the map showing division of the area between the government, Viet Cong, and 'mixed'—in other words an area cleansed of Viet Cong, but not yet under government control.

The officer explains that they have determined six pre-conditions (which he promises to give me in writing) necessary for complete control of the area (establishment of a local council supportive of the government, etc.).

I am interested to see how the officer repeatedly cites the numbers: in the Delta region: 51.5% of the population is under government control, the rest divided equally between the Viet Cong and an 'intermediate state.' Approximately 50% of the Delta area is under Viet Cong control.

We visit the market. It is already late in the morning—9.00 am—and the principal purchases have already been made. But the market still overflows with people and produce: meat, vegetables, and an abundance of all kinds of different, unusual fish. Prices—even of eggs and meat—are not high. One thing is clear—there is no hunger here. Customary meals consist of rice, fish, and vegetables or fruit—all obtainable without any difficulty. Towards evening, during our tour of the refugee's village—I see a man, basket in hand, returning from work; walking around one of the large canals, he throws a primitive net into the water, waits half a minute and starts to pull it in. When he takes the net out, several fish flutter in it. He throws them into his basket and continues in his way, family dinner in hand, without having spent money, effort, or time.

Today's visit is partly by car, partly by helicopter. I am glad to travel by car again. This is the first time that I do not hear firing, and there are signs of normal life all over—civilian traffic on the road (it really does move freely from Saigon to the Cambodian border). Vietnamese guard posts, without Americans, and no sounds of firing.

From the air, the huge Delta region is partly flooded, partly divided in squares. Dull green—rice growing in small plots, from which the green and pale yellow of rice nurseries protrude. After that, the seedlings are transferred to the entire area (in some cases, labor is saved by immediate seeding without the need for nurseries).

We visit several regional villages and farmers' houses. A rich family may have 50–150 dunams of rice: they support a staff of fixed laborers, with more at planting and harvest time. Contract workers receive 50 piasters for planting of 1 dunam (about 1 ½ lirot,[10] or $0.50), and 80–100 piasters for harvesting ($0.75).

Farmers (and merchants of course) who can afford not to sell their produce immediately—place it in storehouses, and charge double the official harvest price—700 piasters for 100 kg, for a total of 1,300 piasters. The season only lasts a few months; the poor—farmers and on-farmers—pay dearly (double the price).

Use of artificial fertilizer is widespread. Farmers have already learned its value and buy it at full price, on the open market.

Most houses in the area contain wide wooden beds, and a table, built the same way as the bed. A coal burner, two or three pots of rice with spices and different side-dishes (meat, fish, vegetables), and a kettle of hot tea boiling all day (tea is cultivated in the vicinity and manufactured locally).

Here and there, there are individual tractors (three in the entire province). Most of the work is done by hand. Plowing is by water buffalo or cows. Produce is transported from the field by wagons with wooden wheels, and boats in the canals which extend through the length of the fields and the 'villages' mostly consist of two rows of houses, on both sides of one canal-length road. The fields are behind the houses.

The farmers are courteous and speak politely and quietly. One of the farmers, Nguyen Van Khoi, who appears to me like an 'economic researcher,' pours tea for us, opens his rice stores, and shows us two coffins with ornate engravings, which he has already bought for his parents—when the time comes.

Most of the farmers draw advances from the merchants and then sell their rice—or at least a part of it—cheaply, because they need the money (at 25% interest). There is a pitiable start-up of cooperatives—3,500 farmers out of a total of about 250,000.

Agrarian reform has apparently taken place in areas under Viet Cong control (they promise me written material on this). From the air, the area seems cultivated and well-organized.

We enter a clean, well-ordered pagoda. The monks recognize who we are and invite us to their homes for tea. This pagoda has fifteen monks, who teach lessons on religion, and live on food donations from the village population, but isolated from them. I ask to see their prayer book: they bring me a book made of dried bamboo, with illustrations of Buddhist deities engraved by hot iron.

We have lunch with the general who both commands and governs this province. When I start talking about the government farm, the eyes of the Vietnamese officials reflect their feelings. They are fully aware of the existing social problems. But that is not to say that all the words, reforms, explanations and ideologies during 20 years of war have not had at least some effect on the population.

During the afternoon, we finally go to visit the refugee villages.

The first village is a disappointment; 250 houses, 20 x 30 meter plots next to each, and that's it. They intend, at some or other time, to take part of the land paid for by leasing fees and give it to the village population. But no one believes a word of this. We fly to the second village—here, they haven't even promised land, rather 3,500 piasters for building and a 6-month supply of food; however, they have not received even that yet. This village was 'created' from farmers who were only 2 kilometers away—in an 'abandoned' area—but were ordered by the government to vacate the area and move here.

This whole business is wretched and miserable. There is also the disappointment of my escorts who, for some reason, assume that I wouldn't get the true picture. The only real human thing during all this time, occurs (through a translator) during a discussion with a (Cambodian) refugee. He speaks enthusiastically (complaining about the government) and frankly: when he sees some sort of beetle entering my ear, he takes it out with his long, agile fingers (filthy from mud)—without thinking that anything is amiss or interrupting his flow of words even for a moment.

The explanations of local officials during the visits make me angry. After all, I know the culture of French colonial rule. Their assumption that the refugees have everything–'what do they need?—They have rice, fish, geese (they live on the water, and receive some rice seeds every evening (to get used to their new home). What do they lack?' But against this depiction of a 'happy man' lacking nothing—I see a man, old before his time, barefoot, clothed in mud to above his thighs, in a reed-and-mud hut (surrounded by endless numbers of children). 'What does he lack?'—This man is—and should be treated like—a human being, not a water buffalo (who roam all over the place, children riding on their backs; when on dry land, they sit on them, but when the buffalo enters the river, the children stand on its back).

The Vietnamese officials in the villages and regions that we visited today tell me that—except for some exchanges of fire—they have not clashed with the Viet Cong for months. They also have encountered no opposition in their bid to broaden their area of control in May, by taking over additional villages. The issues delaying their occupation are twofold:

1. They do not wish to deviate from their plan, complicate and endanger themselves; it is better to 'consolidate' and—next year—take a few more villages. In the meanwhile, they must promise decent services, pave the roads, etc.—to show the villagers the advantages of government rule.
2. Lack of money: a paid militia is conscripted in each village, to protect it. The money received by these guards is important to secure their faith in the government. This year the government is not prepared to expand the number of soldiers from its existing budget. Their policy is therefore—not to attempt new occupations this year, even though (according to what they say) they could, without any difficulty, widen their area of control by 50%.

The Americans must decide the depth of their commitment here. If they really want to 'conquer hearts and minds,' they must offer a better alternative to the farmers than the one put forward by the Viet Cong. The local government is incapable of doing this. To get the job done properly,

American foreign policy must reckon with their remaining here for a long time, to take care of administration, establish model villages and neighborhoods, etc. This may, but will probably not, arouse a wave of sanctions against them, in their role as foreigners. I have been told that the Vietnamese nature is loyalty to the family, not the state. If the American administration takes proper care of this (and there is no reason why not), perhaps the people will prefer it over the Viet Cong.

Jackson says that one of the reasons why the United States doesn't conquer Hanoi—is unwillingness to support their army in the static conquest of a country that—as soon as they leave it—will return to communism. This also explains the mobile nature—so as not to become a static, permanent presence—of their war here, in the South

When I see the poverty-stricken neighborhoods (slums) where the refugees live, it seems to me that, like the French erred 10–20 years by ignoring the national problem (the people's desire for an independent Vietnam), the Americans have made a similar mistake now by ignoring the severe societal problem—because the opponent's ideology is one of communism. Anyone who wants to fight Ho Chi Minh must have another alternative—progressive socialism, not refugee camps, etc. (Asia, Asia).

The villages—especially the refuges villages I saw today—have strengthened my belief, that the Americans have no plan here, nor thoughts of any consequence in the agricultural domain.

At the end of the first day and night without firing, I remember my thoughts before I arrived, about sleeping under a green tree. I am seized by a strong desire to do so today, in the quiet, meticulously kept courtyard of a wealthy Chinese merchant.

Long Xuyên
Saigon, 26 August 1966

We leave for Long Xuyên at 8.15 am with Jackson, first by helicopter, then by car. The quiet, peaceful atmosphere is very attractive.

On the way, we are joined by people occupied in agricultural instruction—two Americans, and a Chinese from Formosa. We visit the house of a farmer, who was the first to grow soya (at American initiative) on one dunam. The soya has already been harvested and crushed, and we see only the seeds. I speak with the farmer, whose name is Thuyet: he cultivates leased land (in return for 25% of the harvest—the fixed, official rate in Vietnam)—50 dunams. One crop of rice, the second of vegetables (rice from November to April, followed by vegetables). He has already grown cucumbers, watermelons and corn. The price he can now get for soya is very good indeed—double that of other vegetables. But this is an artificial price, paid by the United States. The Americans began distributing soya several months ago and only individual farmers were moved by their suggestion. Everything testifies to the fact that the Vietnamese farmer is not conservative. If there is a possibility of getting a better price, he is prepared to use new crops and cultivation methods—fertilizers, etc.—to which he was not previously accustomed.

On the way, I speak with the Chinese instructor from Formosa. A land (garden) unit in Formosa is 3.5 dunams per family, but only because there is no more land. He believes that a family can and must live on 30 dunams (with irrigation, 'garden agriculture').

We continue by car to the Hoa Hao area (a new and separate bodhisattva sect.[11] They were, at first, a secret society, opposed to foreigners, and to control both by the South, and the Communists). The Viet Cong have killed their leader, and now they are sworn enemies. A rich stretch of land with a high standard of living. On both sides of the road, there are substantial houses, cars, and people in city dress.

We have lunch with the district governor, who is also the army commander. He is a landowner (2,500 dunam). Although most of the land is under Viet Cong control, there are many large landholders here.

During lunch, the conversation centers on regional farming and the security situation. Like other officials, he confirms that:

1. Apart from exchange of gunfire, there is quiet and a kind of 'truce' with the Viet Cong.
2. The farmers under Viet Cong control are dissatisfied, and will gladly accept government rule.
3. In order to widen control of the region, a much larger budget would be required, to allow conscription of more soldiers (from amongst interested villagers).

With control of additional villages—he would be able to conscript the necessary soldiers from these as well, to implement government development: roads, schools, a clinic, refuge relief, etc.

For 'pacification,' he wants neither American soldiers, nor the regular Vietnamese Army who just help themselves to chickens and women (and are of no use at all). He wants money, arms and supplies, then he will conscript soldiers from amongst the locals.

The issue of local conscription is a central demand by the local rulers, and it seems to me that they have turned it into bribery and blackmail: give us more money—and we will intensify the war; if you don't, the situation remain the same.

There are three advantages to local conscription:

1. Those who are conscripted are under control of the local governor, not the central government.
2. It will be a significant economic boost for the people in the district—salaries (they continue their farm work), uniforms, foodstuffs, etc.
3. No antagonism to a foreign army would be created which interferes with community life and those who receive the assistance.

In any event, this year they do not plan to renew or widen their activities. Before this, they want to become established (complete developmental activities already begun, extend governmental organization, etc.). Next year—if they receive more money and conscript more men, and if the American-Vietnamese Army extends the war to the Vietnam-Cambodia border area, and if the Viet Cong are pushed back into the jungle—then they will control more villages. They speak of ending the war in 1971.

We return to Saigon by helicopter. I attend the daily press reception. General DePuy gives me the report. His men have just fought a hard battle with the enemy. One of his patrols found itself surrounded by a 400 men-strong Viet Cong battalion. The battle lasted about 2 days, and DePuy brought up 2 battalions of reinforcements, armor, and air and artillery assistance. More than fifty Viet Cong were killed (perhaps more by the artillery, but an inspection has not yet been made in the jungle). The Americans had more than thirty killed and about a hundred and eighty wounded. Many of the American killed and wounded were due to friendly fire by two napalm bombs.

The atmosphere is somber: not just because of the napalm, but also the fact that, despite all material advantages, they failed, after 2 days, to 'chop the enemy up into little pieces.' I want to go up to DePuy, but don't have the heart to do so. The mood is grim.

I have dinner with the French consul general, who wants to hear my impressions. In answer to my question, he said that the FLN (National Liberation Front of Algeria[12]) office in Cambodia could arrange my communication with the Viet Cong leadership. He suggests sending a telegram to the French consul there (but only after De Gaulle leaves on 3 September). If there are no changes in my plans—I will travel to Cambodia and try to contact the Viet Cong there.

According to the district governor of Hoa Hao [*sic*][13], agrarian reform has been introduced in the Viet Cong areas. Each farmer is given about 10 dunams, and the amount of rice necessary for a family is reckoned according to the number of their children, etc. On average, about a ton of rice per year (3 kg per day, for a small family). The calculation is exact, even including details like haircuts, which are paid for in rice.

'Winning of hearts and minds' in Vietnam must depend on three things:

1. Existing social injustices—a way must be found to correct them.
2. The refugees;
3. The ideological struggle with communism, which has gone on here for the past 20 years. Social factors must take a prominent place in the war against Ho Chi Minh.

Both the problems here, and their severity, must be properly acknowledged, before the above three issues can be properly addressed.

This is my last evening in Saigon (Vietnam), with the background of B-52 bombings.

The End of My Visit
27 August 1966
Saigon

The Prime Minister, General Ky, is unable to see me, there are no diplomatic relations with Israel (Israel refuses).

At 11.00 am a lengthy discussion (followed by a lunch of the best cuisine the country has to offer) with the chief of staff of the Vietnamese Army and the minister of defense (and deputy prime minister) of the South Vietnamese government, General Nguyen Van Thieu.[14]

The significance of the conversation—at least from my point of view—is that I give my main conclusions about the war, about which these generals are experts. The chief of staff is intelligent, clear and precise in his definition and analysis of military policy.

He says that there are objective difficulties related to the way of life, etc. in areas controlled by the Viet Cong. There is increasing pressure on the population. A few years ago, they were still exempt from taxes. But after that, a 25% tax was introduced, and, when conditions worsened, a 50% tax. Tax is currently 70% of the rice harvest. Vietnamese soldiers obtain 700 gm of rice per day (apart from other nutritional materials). The same applies to conscripts. In the beginning, only volunteers were conscripted. After that, conscription became obligatory and is now from 15–16 upwards. From this point of view, there is no doubt that life in Viet Cong controlled areas is much more difficult than that under government control. Nevertheless, he cannot explain to me why the numbers of Viet Cong deserters are so small.

Regarding the future of the war, they both speak admiringly of General Giap[15]; he is the right man to compel the Hanoi government to stop the war. Giap is thoroughly informed about conditions at the front, and Viet Cong defeats at the hands of the Americans. They see the turning point—or first phase of changing campaign alignment—as beginning at the start of the New Year:

1. After the upcoming Vietnamese elections, the North will despair of the government collapsing.
2. After the November elections (in the United States), they will despair of President Johnson's strategy collapsing.
3. The Viet Cong will have to renew the fighting during the dry season (January to June), and will be either beaten, or will have to find another alternative.

Although they hope that the alternative will be readiness for a peace conference, they assume that it will not occur; there will rather be a change to guerilla warfare. This has nothing to do with any specific organizational issues: the Viet Cong have only recently tried to organize themselves into larger than battalion-size units—regiments and divisions. Before that, they fought in small guerilla groups, to which they could easily return.

They think that the Viet Cong are doing better in the Delta than in the mountains, which are thinly populated with mountain tribes, who have less food. By contrast, they have an abundant food supply in the

Delta, and a civilian population that supplies them with good intelligence, opportunities for reconnaissance, deception and conscription. Southern forces have no assistance in the North, and do not know the area well.

Their children go to French, not American, schools, and French is still the ruling language of culture (I promise to send a French translation of my Sinai book[16] to the army chief of staff).

During last night's conversation with the French consul, he asks (distrustfully) whether the Viet Cong are really a regular army. I answer yes, although in primitive form.

The possibility for guerilla warfare here is very great. For example, for 100 piasters I was allowed to pass through the security gate, with a civilian driver. I assume that a local guerilla representative can do this for even less.

I leave Saigon for Bangkok at 5.00 pm.[17]

Endnotes

1 Chapter 1, note 47.

2 Chapter 1, note 50.

3 United States Agency for International Development (USAID)

4 Chapter 1, note 57.

5 Chapter 4, note 5.

6 Chapter 1, note 10.

7 Chapter 5, note 39.

8 It is unclear what exactly Dayan is referring to here. It may be the South Vietnamese Regional Forces, originally the Civil Guard, that was a component of the Army of the Republic of Vietnam (ARVN).

9 Four dunams is one acre.

10 Old Israeli currency.

11 Hòa Hảo is a Vietnamese new religious movement. It is described either as a syncretistic folk religion or as a sect of Buddhism.

12 The *Front de libération nationale*) commonly known by its French acronym FLN, was the principal nationalist movement during the Algerian War and the sole legal and ruling political party of the Algerian state until other parties were legalized in 1989.

13 Dayan is likely referring to either Bien Hoa or Kien Hoa, both near Saigon. Hòa Hảo is a Buddhist religious movement that emerged in the Mekong Delta in 1939. The movement was very strong around Saigon and further south during the period of Dayan's visit.

14 Nguyễn Văn Thiệu (1923–2001), South Vietnamese officer and politician who was president of South Vietnam from 1967–1975.

15 Chapter 5, note 27.

16 Published in 1966 in French, *Journal de la Campagne du Sinaï 1956* (English, *Diary of the Sinai Campaign 1965*) is a personal account of the Sinai Campaign of 1956, also known as the Suez Crisis.

17 As will be seen in the next chapter, Dayan's plans to visit Cambodia after Vietnam changed.

7

HOMEWARD BOUND

Bangkok
30 August 1966

Yesterday, I met two central Thai army figures. I was invited to lunch by the chief of intelligence—General Sankadis,[1] after that I met with the army chief of staff—who is also the deputy minister of defense, Air Vice Marshall Daval.[2]

Conversation centered around their opinions on the link between the war in Vietnam and the regional situation, and their long-term thoughts on the struggle between China and the United States, or between communism and the West.

Quite obviously both are violently anti-Communist, and against China's wish to extend its domain through local Communist parties. Their main points:

1. The situation has improved since the Americans entered the war—Communist pressure has diminished.

2. The Thai Communist party started 'working their wonders' in Northern Thailand some time ago. They have begun to take care of villages—medical, humanitarian, and other economic assistance—and to incite the people against the government in Bangkok. They tell the villagers that in Bangkok people are rolling in money, which should be taken from them and redistributed amongst the farmers.

3. Tiny Laos, which at first opposed American bombing in their area (against North Vietnamese infiltration), has gradually agreed to this—at first only with machine guns, but later with bombs and now with napalm.[3]

4. The chief of staff is of the opinion that America must take Hanoi and be prepared for a prolonged war lasting years.

5. The chief of intelligence disagrees quite strongly with a long-term American presence here. He even says that, if the question of influence of China or the United States arises, China is Thailand's northern neighbor and is nearer to them that the United States.

The chief of staff has a lot to say about Burma.[4] Ne Win,[5] and another Burmese general who is presently under house arrest, were, at first, good friends—but Ne Win has gone his own way, from which there is no return. He has good intentions: to bring progress to his country and protect neutrality between China and the West. The Thai chief of staff finds a degree of justification in what Ne Win has done, because of the long Sino-Burmese border, which forces Burma to recognize China and preserve good relations. But Ne Win's methods, which prevent any opposition (by imprisonment, exile, ousting his opponents), together with terrible bureaucracy (nationalization, and transfer of life into the hands of government officials)—is destroying Burma. At the end of the day, Ne Win will either be hanged by others or will have to hang himself.

The chief of staff says that even Chou En Lai[6] has told Ne Win that his method of leadership (government bureaucracy controlling, overseeing, and directing everything) is more extreme than communism (Ne Win admitted that to the Thai chief of staff).

It seems to me that two men's optimism is a little 'for export,' but that, in their hearts, they fear entanglement in current regional events, and China's possible entrance into the war. They expect the war to be prolonged, and do not look favorably on long-term American bases on

Thai soil. But they clearly understand the current situation—either the Americans fight and remain here, or the Communists—in North Vietnam, even in Thailand and Laos—will conquer area after area. In such a case, they prefer the Americans.

At my request, we pay a visit to the American ambassador. He has mentioned my trip to Washington, and said that they value my opinion, because of my reputation as a soldier. After that, we visit the Burmese ambassador to ensure that he has told his government in advance that I am arriving in Burma on Wednesday (30 August).

In the meanwhile, I receive a message via the American embassy here, that McNamara and Johnson—but not Rusk[7]—promise meetings with me, and request conformation of when I will come to Washington. I say yes, during the week of 11–17 September.

30 August 1966
Bangkok

At 9.30 am, meeting with staff of the regional UN organization involved with regional planning. They have a 5-year budget of 3 million dollars for coordination, etc.: no small sum of money. The body itself neither operates nor finances, only coordinates, through mediation and promotion, etc. Despite our request, Israel has not been accepted into this organization: U Thant[8] does not want Israeli-Arab issues to complicate its principal projects: a Mekong River plan, a trans-Asian highway, exploitation of mineral wealth, opening of industry, etc. At the end of the day the plan, and its implementation, depend on each individual country. But the mere fact of its existence—with assistance from the Bank of Asia when it begins operating—will strengthen all the organization's projects.

The Mekong Committee is a short- and long-term plan—one project, in each country through which the river flows: Cambodia, Thailand, Laos and Vietnam.[9] Electricity production, accumulation of water for irrigation. A small amount of work has already been completed, and the project continues.

These large plans, costing hundreds of millions of dollars (4 at first, a total of about 100 for each country, especially Cambodia)—are still in the clarification and planning stages. There is clearly a great desire to consummate this huge agricultural development project. Irrigation and electricity production for industry.

Lunch at the ambassador's home—'males only.' A lively discussion about the war in Vietnam. At 3.00 pm I receive a briefing from an American colonel at his headquarters on American activities and arrangements here. There are about 25,000 American soldiers in Thailand. I am not impressed that they place any emphasis on training of the Thai army.

Meeting with General Minh[10] at 4.30 pm—as I have said before, he is in exile here. He speaks slow, faltering English, but his sincere personality is very impressive. He sees himself as a Vietnamese patriot, who cannot return to his country, to help solve its problems; he has tried to return, but his plane was not allowed to land—General Ky[11] fears his popularity inside the army. Minh maintains that the Americans have taken on too much of the fighting. The Vietnamese Army should be the main fighting force, with Americans just there to assist them. He does not believe that the North can be forced to the bargaining table: they are able to live at a primitive level, without roads, electricity, etc.

His chief criticism—which he voices in a very restrained and heartfelt way—is that instead of war being between the people (soldiers, etc.) of South Vietnam against the North and the Viet Cong—they are attempting to rule by force, with American assistance. This major American role in the war leads to many negative consequences:

1. Lack of confidence in Vietnamese leadership. (He says that he himself has no confidence in it—its main commanders, whom he doesn't want to name, are smuggling money out of the country, on the supposition that they will, sooner or later, have to leave Vietnam and live overseas).

2. Vietnamese (in and out of the military) cannot compete with Americans in money, etc. Prices have risen steeply—leaving the locals in a difficult position (resulting in prostitution—young women must do something, to feed their families).

The general believes that now is the great opportunity to change the Vietnamese population, currently under Viet Cong rule, into supporters of the government. Viet Cong rule is very heavy-handed. The people themselves will not rebel if the Viet Cong depart from the villages; just the opposite—the farmers will support the government, on condition that it has a genuine administration. But time after time the Americans harm the villagers through their war with the Viet Cong, by bombing, etc.—This makes the population susceptible to anti-American Viet Cong indoctrination: Americans are portrayed as foreigners who want to rule with the aid of traitors from within.

He speaks very highly of Chinese intelligence (Formosa)—they are loyal to one another. There are many Chinese in every country and every region in this part of the world, and 'Chinese only speak, and transmit information, to other Chinese.' There is no amity between the populations of the South and the North. Northerners 'look down their noses' at Southerners and try to give them orders. As above, there is also the relationship between all Vietnamese and the 'Red' Chinese. All of us (D. Ben Ari, Chazi Carmel[12] and myself) are impressed by the sincerity and pain in what he says about his homeland. It is torn asunder by war and loss of independence, and its leaders do not believe in themselves, but rather in others.

He also believes that the Viet Cong will go over to future guerilla warfare—not because they have given up on regular war, but because Communist theoretical guidelines see guerilla warfare as an early phase of a final war, which must be continued, if the war cannot be ended. Radio Hanoi broadcasts all sorts of lies (about fictitious victories) but also emphasizes guerilla warfare and terror.

At 7.00 pm, I meet with the Israeli instructors of local model farms. They are introducing irrigation, to achieve two crops. They grow cotton, sorghum, and corn, and want to introduce citrus as well as cooperative marketing (Thais pay Chinese merchants 100% interest). They say that the Thais are very good farmers and praise their agricultural university.

Dinner with the minister of information.

My general impression: even those who do not accept communism (these are the only ones I meet) also do not accept an American alternative. The most they want is assistance, but not large numbers of American bases here for the foreseeable future. They are foreigners, and harm Thai independence—which is extremely important because Thais regard themselves with pride, not as inferiors.[13]

In Thai eyes, Europeans are not superior to them in any way (perhaps the opposite). But, at the same time, they are not hostile, because they are happy to see them as a source of money.

When I tell General Minh that in the North the rice ration is tiny, he replies: 'yes, but each one gets an equal share.—That is the main thing.'

The American assertion that their activities in Vietnam have curbed the spread of communism has no support in Thailand. The locals are not convinced that, were it not for Vietnam, the position of communism here would be strengthened. Indeed, the opposite cannot be excluded—The Americans serve as grist to the Communist mill.

31 August 1966
Bangkok

During my stay here, I have been feeling unwell. This has not influenced my meeting schedule, but I cannot concentrate on my writing: when I finally get to my hotel room, I am already too tired to do anything but sleep.

This morning, we (the ambassador, military attaché and I) visit General Prapat,[14] and after that the minister of development—Sarasin.[15] After the prime minister, they are considered the central figures in the government. General Prapat is the deputy prime minister, minister of the interior, and commander of Thai land forces. He is a decisive man, and future candidate for prime minister. He opposes parliamentary rule and wants a one-party state. He says that America has invested about 2 billion dollars in Thailand over the years (including projects still under way: airports, roads, and harbors). At a rate of 200 million dollars each year in current allocation of funds—apart from special investments—they have no enmity whatsoever against the United States. But the Communists do not exploit this for internal incitement. Were it not for the Americans, communism would spread, not by a Chinese war of expansion, but by local, regional rebellions, with financial and arms support by China, the Soviet Union, Ho Chi Minh, etc. This was—and to a certain extent still is—the situation in the strip of North-Eastern Thailand (10 million of the country's poorest inhabitants), near the pro-Communist region of Laos, on the other side of the Mekong River boundary.

The only criticisms they voice about Americans are:

1. They do not develop local forces properly, preferring instead to erect their own logistical bases.
2. Their assistance is incomplete and unplanned, but rather improvised and made up of items (arms, etc.) which the Americans have in plenty, but Thailand doesn't need.

As for the war with Vietnam—they believe that the Americans must develop more local forces there as well (with less Americanization of the war) to take Hanoi, and that China will not enter the war if this happens. But the United States is not doing this, because it does not want to appear to the world as an aggressor. The minister of development—Sarasin—is a pleasant man, well-liked by the public (according to our ambassador). An outstandingly complete map-room (more organized than any war room I have seen previously): portable maps, diagrams, portfolios, etc.—on every conceivable subject. Their annual developmental budget is about 200 million dollars.

Half is for roads and irrigation; widening of irrigation areas is the main agricultural development project (the goal is to achieve two crops a year, with the second not necessarily rice). Farming is the most important developmental issue. Thailand still exports rice, despite have doubled its population.

1. Unlike its neighbors, Thailand has never been under foreign or colonial occupation. This is certainly one of the reasons they do not recoil from an American presence. Additionally, the Americans really do give (money, roads, work, etc.) and do not take. They also do not interfere with communal life as much as in Vietnam. There are a total of 25,000 soldiers here, as opposed to half a million in Vietnam;
2. American influence is non-colonial and non-exploitative.
3. All Anti-Communists know that, were it not for the Americans, communism would already have taken over.
4. The Thai military and civilian government elite derive material benefit from American development, financial and otherwise.

'Asian communism' has different forms, with different effects:

1. Ideologically: Even the 'great General Minh' speaks highly (even more than that) of the leaders of Vietnamese Communism (Ho Chi Minh and Giap[16]), and their achievements; 'In North Vietnam each person has the same 140 g of rice, therefore they can put up with it.'
2. Large areas and populations of Thailand—especially the poorest and most distant parts—are currently under Viet Cong rule: the Thais call it 'Minh Kong,' in other words, the party of Ho Chi Minh. Added to that: a portion of Laos (Pathet Lao, one million), North Vietnam, and of Burma, etc. Therefore, communism has become part of the non-Communist Asian way of life, because it exists within the non-Communist body politic.

3. The Chinese phenomenon: Chinese in Indochina, Thailand, etc. are the merchants, and biggest exploiters of the people. But they are Chinese with Thai citizenship, some of them with ties to (Red) China; some have been accepted—if they wish it—for study in Beijing, and many do it. The Chinese do not sever contact with their people, mother tongue, or nation, and constitute a helping hand for China.

4. There is respect, fear, and solidarity with China against the West. After all, China is Asian, and nearer to the hearts of Asians than America.

The problem of social prestige (in Thailand and Vietnam) is not as severe in village areas, as it is in the faraway cities, which are strange and non-comparable. There is no hunger or lack of food (rice, fruit, fish and vegetables) or land. It is easy to prepare more land for cultivation and irrigation from the jungle.

Here, as in Africa, if there are complaints and opposition, their sources are cultural, ideological, and nationalistic, in other words from the urban student community, etc. Social prestige is not a motive for opposition to the West.

Traditional Thai religious and national tolerance is much greater than that seen in Middle East Moslems. Trade—the ability to earn—determine and advance everything. Additionally, Americans are not the 'only players' here. Merchants from Germany, Italy and other countries who become rich, also inject money into the economy, and none of this is interpreted as exploitation.

Rangoon[17]
31 August 1966

Upon arrival, I see that the people from the embassy who greet me looking depressed. Our ambassador (located in Cambodia) left instructions to treat me as they would a 'senior journalist'—and that is all. We turn to the Department of Protocol of the Burmese Department of Information, but they tell me that, because I had contacted them at the last minute, they could not arrange meetings, etc. Perhaps the Deputy Director of the Department of Information could see me at some or other time. I tell them not to worry. I am not expecting them to arrange important meetings for me here: I will try to do so myself. If I succeed—good. If not—at least I will have rested for a few days.

We walk around in the city a bit. We visit the synagogue—large, beautifully decorated, but empty. From more than 500 Jews living in Rangoon, less than ten families remain. In the market— large amounts of fruit, vegetables, etc. There is no hunger here. Long lines at the bus terminals (there are very few cars). We visit a large pagoda—crowds praying normally, no end of flowers, and majestic gold and other decorations. Religion has not weakened here, and there is no contradiction between it and the regime—Burmese communism.

Towards evening, I return to my hotel. I sit in the lounge and drink tea. At that same time, the Indonesian Embassy is holding a cocktail party in honor of their National Day. The invitees pass through the lounge in which I am sitting on their way to the cocktail hall. Many (locals as well as representatives from foreign embassies) recognize me, and a few come up to greet me. They are followed by representatives of Reuter and Agence-France Presse, who ask why I am here. I tell them that I want to meet with the men in charge, see the country, understand their approach to social and political issues, and write about it. If they don't want to meet with me—I will leave without any meetings and write about that as well. I mention that I have come here on the recommendation of U Thant. They promise to clarify the possibility of meetings with local leaders and get back to me with any information. At night (after dinner at the house of the Israeli economic attaché,—Zvi Timor) one of the journalists (from Agence-France Presse) calls and says that we should meet at the hotel at 8.00 am tomorrow morning.

The hotel has 400 rooms but is almost completely empty—less than ten guests.

1 September 1966
Rangoon

My attempts to meet with someone senior from the army, agricultural, economic, or information departments—external and internal—do not go well. All my telephonic or personal requests are answered courteously but referred to the Department of External Affairs.

At noon, I am invited (for some reason) to meet with the director of the broadcasting ministry. Four men wait at the entrance and greeted me courteously, with shining faces.

We enter a large meeting room. After a few moments, a young army major in uniform arrives. Like all other institutions, the broadcasting ministry is controlled by the army: its workers are under the major's direction. He too is very courteous, and orders tea and biscuits, He even tries to arrange a meeting for me (after he hears what I want), but returns shame-facedly and tells me that I must first report to the Department of External Affairs (we have already tried there via the embassy but I was not even given the courtesy of a telephone conversation: they were all busy at meetings, or not in the office ...).

I do not get the impression that the major is upset that he doesn't have the authority to arrange something for me. Perhaps the opposite is true: perhaps contact with foreigners (as with others) is a burden, which he sees as a kind of tax that he is sometimes obliged to pay (he repeats several times that Israel is a friendly country, and that he therefore really wants to help me get what I want). But, in retrospect, if he were freed from this burden and someone else took it on, no one would be more pleased than he. The Department of External Affairs should take care of this, allowing him to return to his assignments, his language, his people and things that he knows well, and not have to deal with difficult issues, such as my language, and my own problems. The truth is that I can't even get angry. It may be that I see my personal reflection in his public behavior: from time to time, when I have guests and am not in a good mood to receive them—I escape to my collection of (ancient) vases and take refuge behind (pretending to examine) my broken potsherds.

The most impressive thing about the morning visit and the discussion that takes place, is the civilian subordination to the young officer. The officer speaks and behaves aggressively to them (he permits tea to be brought to them as well ...). His entire appearance radiates superiority—the military! The ruling authority! The civilians—who are certainly knowledgeable and experienced officials with their own portfolios—obediently tag along behind him. The saddest thing in today's tour—is the stagnation of daily life. As already said, ten hotel guests out of 400 rooms, and only six passengers on the plane flying here. I register for a flight to Namsang in the country's hinterland and ask how many passengers have already booked—I am the only one. I don't have to wait in line at the post office: crowds of clerks—but only a few buy stamps. I go into the postmaster's office—he sits reading a newspaper.

Stores are forlorn and empty, cupboards and shelves completely bare. In the special stores for diplomats, a varying assortment of all sorts of canned goods, confiscated from a variety of (nationalized) grocery stores at the time of nationalization: who knows when they were produced. Tinned meats with Arabic letters (? Pakistan), sausage, etc.—as many different sizes, shapes and forms as can be found in a food museum. Rice is provided in small newspaper bags that the shopkeeper folds on the spot—and this is a special store for diplomats.

The market is silent. The currency is worth ¼ of its true value (on the black market). Workers are not interested in productivity—they get a (low) daily wage, without merit enhancement. The managers (all members of the military) know nothing and are not interested in speeding up production. The result—a kind of distribution of most essentials (free rice and vegetables, but no potatoes, etc.). There are organizational and material problems in distribution, for example of soap and toothpaste between villagers and city-dwellers. The result—is a black market. But—there is no hunger, and it is possible to live, although at a low standard. Certainly one position—the military ruling class—receives special benefits (cars, spacious apartments, clothes, etc.). This is so obvious that it is infuriating, but no one predicts a rebellion from below. If a rebellion does occur, it will be a putsch by another group of officers, against the current rulers. In the meanwhile, General Ne Win—'Chairman of the Revolutionary Council'—does not fear travelling out of the country for long periods several times a year, with resultant loss of control in his absence. What is happening in other countries is not hidden from the people—Thailand: Bangkok, which blossoms and shines with plenty, Cambodia, etc.—But in the past, they know that places like Paris and London exist as well. The feeling here—let us live as we wish, without foreigners—neither from outside, nor from inside—(get rid of the Burmese Indians, foreign delegations, etc.).

It is as if the historical concept of the 'Burmese way' has been given a new symbolic meaning. We do not want foreign influence, a foreign way of life, and foreign culture—we want to be ourselves—Burmese. I hear this repeated explanation from everyone.

Ne Win has travelled to Washington and London, to demonstrate his independence from Beijing and Moscow. Burma's relations with Western countries are not dictated by China and occur without previous Chinese consultation. Relations with China are organized and honorable. China desires a respectful neighbor who will not cause any problems—and Burma cooperates. There is no loathing or fear that China will gain control of Burmese regions. Both countries have reached a border arrangement, and both recognize and respect it.

It is possible that Burma might change its economic structure, weaken totalitarian rule on all parts of the market—even allowing small stores—and allow some degree of liberalization. But they would only do this if they were convinced that this is in the country's interests; then, ideological considerations would not delay reforms. But in the meanwhile, Burma aggressively and stubbornly refuses to participate in international activities—whether regional, or influenced by foreign elements.

For example 'the Asian highway' or any other developmental projects that receive American money. They receive both compensation and loans from Japan—30 million dollars—but spend the money only for investment, not for need.

Freedom of the press and freedom of speech in Burma are remarkable (the first page of today's newspaper is full of local news about rebel bombing of a railway, misappropriations and other failures. I am told that sometimes they even write about entire storehouses of goods not in the market, because their contents have been thrown into the sea. For some reason, the merchandise had rotted in storehouses and could not be brought to market).

There are even angry letters to the editor criticizing the country's economic policies—There is freedom of the press, and the government does nothing to stop it.

There is also no contradiction between church and state. Even the elite are devout Buddhists. Pagodas are full, and there is no propaganda against Buddhism (or any other religion). However, many Christian priests have been expelled, as part of the campaign against foreign influence—the mission here that administered schools, etc. (By the way, Burma has changed to instruction in the Burmese language, instead of English, as was the custom until now). Steps have also been taken against the influence of Buddhist monks and their interference with political life, and they now limit themselves to religious issues.

Economic grievances do exist, but they will certainly not turn into a rebellion while there is no shortage of essential foodstuffs—rice, etc. These are not rationed, and the population have what they need. Only exports have decreased, and there is a lack of foreign currency.

In summary, the economic situation is as follows:

1. Total and complete nationalization, and subjugation by a military that has no professional experience, etc. and abolition of everything that encourages production and trade. The economy is

silenced leading to shortages of most consumer goods. But there are enough essential items, such as foodstuffs and clothing.

2. Introversion—the desire to live an independent life, avoiding any foreign cultural or economic pressure or influence from the West or from India ('we are putting our house in order now and not accepting foreigners'). This attitude is persistent and stubborn and does not create a people envious of rich Thailand, or of Americans influence. They prefer to renounce the plenty, just not to be harmed by its influence.

3. 'Communism.'—This is not true communism, but rather a kind of pragmatic socialism, in the face of a neutral foreign policy. Ties of respect and loyalty to China.—But no Chinese dictates; China has no interest in this anyway.

4. Military rule—as a reaction against internal corruption, and to ensure achievement of goals—This is accompanied by nationalization, removal of the previous administrators and administration, etc. But it has depleted the military, who have taken on new responsibilities. Economic enterprises taken over by the military have been weakened even more (even my hotel manager is an army officer); the military has neither the knowledge nor ability to give the workers incentives to increase production.

5. People's reaction—shortages, criticism, but no signs of rebellion.

Something more about the country's government: Various areas (about 25% of the population) are not really under government rule. In other words: Rangoon can send in the army and enter wherever they wish, but when they return—unruliness, and refusal to recognize the central government, also return. There are several causes for this: Several tribes desire autonomy.[18] There is also a real Communist opposition, who want a change to true Communist rule. Various other 'movements' have degenerated into groups of bandits. The government—in other words the army—want to overcome them, by (as I heard from the French embassy) obtaining a couple of dozen helicopters, to organize a mobile unit (battalion) that can rapidly reach anywhere and create order in areas of unruliness. (The Americans have not succeeded in introducing their 'way of life,' but at least the 'cavalry' 'search and destroy' concepts have penetrated).

There is no real connection between the four basic phenomena (military rule; nationalization and government control; economic pauperization; lack of secure rule over wide parts of the country). But, taken together, they create a very negative picture:

Rule by government elite—which will get worse—creates a layer of 'protektsiya,' corruption, etc. Rule by nationalization and bureaucratic totalitarianism has silenced all initiative and desire to increase production, and there is neither knowledge, nor ability in the direction of, enterprises, trade and administration.

Economic pauperization also results from a policy of isolationism—not to participate in international enterprises, allow foreign capital to come in, or accept developmental assistance from the West—Independent and closed 'Burmism'—in the background of a developing and flourishing world all around them.

Other countries, for example Thailand, also have a degree of governmental powerlessness and lack of security. But in Burma the following is added: the huge burden that rests on the army administering the country, preventing it from developing and implementing its regular assignments. This is all fertile ground for opposition to the central government, because of the economic ebb.

If Burma's economic and planning structure is not changed in time—the situation will surely reach a breaking point.

Parallel manifestations in areas of Vietnam, Cambodia, Thailand and Burma which are not under government control, are of interest. The trends are different, but the causes the same: weak governments without good administration, difficulties with transportation, and contact with distant, cut-off parts of the country, a tribal tendency, Communist influence.

Despite the many similarities between Africa and Asia (easy life, primitiveness of farmers, small jungle plots turned into areas that sustain life, etc.)—one basic difference stands out: In developed levels of African society, there is a feeling of inferiority, compared to Western culture: they have nothing to stand against it. Not religion, writing, literature, history, nor an ethical way of life. In Asia things are different—Buddhism, literature, their writings and way of life (including building, the arts, dance, etc.)—They have a foundation of spiritual independence. Some have been educated in the West, and now they want their own Asian national independence.

In the evening: first, at the Britain-Burma Society. An evening of theater, singing, and performance (by local amateurs). The chairman is an Indian professor. The hall is full of people from Britain, Indians, and others, but almost no Burmese.

After that, a farewell party for the British military attaché, Colonel Baines (I meet him in the morning, and he explains the position here to me). One of the guests—an Indian lady, performs a dance of her own, with taste, delicacy, and great artistic feeling.

At dinner, I sit next to a respected Burmese lady: Daw (honorific for Mrs.) Khin-Khin. Her late husband, a general, was the first Burmese to be knighted by the British. She herself initiates contact with me (asking whether I know her brother-in-law Captain Barber—I really do know him, from Tel Aviv).[19] She pours her heart out to me about conditions here. The Americans have invited fifteen Burmese to the party—but only one comes, and he only to inform the government about who is there, etc. She tells me that a notice has gone out forbidding meeting and speaking with foreigners. If this is done the contents of the conversation must be reported to the government. The wife of a Sikh complains about the advantages taken by Burmese officers, the prohibition against leaving the country, flourishing black market, lack of proper trade, etc. It really breaks my heart to hear—'I am not prepared to write a report about every conversation I have with a foreigner. I have never informed, and will certainly not start doing so at my age.' In connection with the government's 'new method'—she says: 'in the past Burmese put silk kerchiefs on their heads—denoting their status as *thakin*[20]—but now they must put on farmer's kerchiefs ...' .There is no doubt that this society of former aristocrats, with trade, economic and cultural links with the West, do not toe the current party line.

The most difficult thing to hear from them, is the demand to report on all conversations with foreigners to the authorities.

In the hotel, I find a bill for $1.50 for 'a special order of fruit in my room'—This must have been for half a pomelo on my room table. Perhaps it is because, during the earlier part of the evening, I had brought the pomelo down to reception. I tell them that I couldn't care about the $1.50, but that it is a crime to charge such a price for it (half a pomelo + service + two kinds of taxes + percentages on all the above). I leave them the half pomelo as a gift.

With all my desire to understand them as best I can, it is impossible to hide from the terribly negative aspects of what is being done here. Mainly: revocation of individual freedom, isolationism, and treatment of foreigners as potential enemies.

One of my old friends, Aung Gyi, a previous director of the Department of Security and Ne Win's deputy,[21] is now under house arrest because of his opposition to the present government. They told me in Bangkok that, when Ne Win first suggested his program of military rule, he said that, in 5 years, everything in the country would be corrected, and the people could return to their former lives of freedom. Aung Gyi replied that he could add a zero to the 5, making it 50, and the plan still wouldn't work. The conversation ended with Aung Gyi's imprisonment.

In 1958, the army took over the government as 'trustees' for a period of 2 years, to fight against corruption. In 1960—it returned the country to civilian rule, but in 1962 it took control again—this time, permanently.[22]

2 September 1966
Rangoon

In the morning, I walk around the city and its environs. It is pouring with rain—and the car's air conditioning must be turned on (to complete the picture, on the side of the road children are swarming around, in muddy puddles). The road looks forlorn and depressing. There is no importation of colored dye; the result—there has been no colored paint for years, in houses, on walls, etc. (except for pagodas and military buildings—the new 'temples'). Lack of new coats of paint not only damages the houses' appearances but also their impermeability to water, and mosses rot all over.

The buses look like those used decades ago and are crammed inside, faded and crumpled outside (sometimes a bus passes, and behind it, people hang on the roof ladder—'travelling' 40 miles to the next station).

I have lunch with the French ambassador and a few other guests. Topic of conversation—the war in Vietnam.

At 3.00 pm, meeting with the person in charge of propaganda, Mior Hon Kin[23] a pleasant, seemingly experienced young man who laughs a lot. He is not serving in the army now. He was in the navy, was discharged, and now is a civilian.

In answer to my question:—Burma refuses to participate in international development projects, if this obliges her to enter a political framework (Asian Bank). But in other cases, without political motives, Burma is interested in participating within the framework of international development. Burma does not participate in the building of the 'friendship road,' because each country pays for paving of its own roads, and it has more pressing financial and development needs.

They are planning to overcome shortages of essential consumer goods by weakening government supervision, allowing farmers (90%) to leave a certain amount for themselves (which they can, of course, sell on the black market, which he doesn't mention by name). They will be able to sell wherever they wish, but at fixed prices (in reality, also permission to sell at inflated prices).

He knows that the army today has a 'preferred status,' with advantages over others, which obviously creates negative feelings. 'The general' is aware of this, and has said: 'every man, including the army, is capable of development; therefore, the army must promise to avoid taking advantage.' Within 6 months, a party council, with a civilian deciding majority, will be the ruling body.

I am interested to hear that, in all his answers to my questions, he does not once assert that conditions are good. Instead, he confirms my criticisms and says that they intend to correct matters. One thing will not be permitted though: they will not permit return to a free market economy.

At the conclusion of our conversation, I tell him that I want to travel to Namsang tomorrow. 'Certainly, certainly,' he answers. I ask for the name of the man responsible for the Namsang area—(he looks in all the books, asks by telephone, but cannot find anything (he promises to call me at the hotel, if he finds something).

After the conversation, we go to buy my plane ticket. After several searches and explanations,—the director says that he is not able to sell me a plane ticket if I don't have permission. I ask to see his written instructions on this; he takes out a book (written in Burmese), and reads out that all foreign embassy employees wishing to fly into the hinterland, require a permit from the Department of External Affairs. Tourists need a permit from the Department of Internal Affairs. That is the only difference.

Zvi Timor and I return with the director of propaganda, and at his invitation, we travel together to the airline (UBA). After speaking with the director and making several telephone calls, he finally shrugs his shoulders and informs me that, indeed, without a permit—which I would not receive in the short time left to me—it is impossible for me to travel inside Burma.

So ends one plan.

On the way back, we go into a store to buy a periodical—*Time* (by the way, both *Time* and *Newsweek* can be found in all government offices). As it turns out, the following are needed to buy a periodical:

1. A questionnaire must be filled out: name of newspaper, price, buyer, number of copies, etc.
2. Stand in line beside a small hatch, and pay. The cashier has no change (the farmers hoard coins, because they don't believe in paper money)—and we don't have small change.
3. After this has been sorted out, with three forms in hand—two copies of proof of payment to the seller, one for ourselves—we receive the periodical.

In the office, Zvi Timor shows me correspondence between himself and various Burmese government departments: he had requested permission from the highest authority to export some varieties of corn seeds (30 g per variety) and garlic (The letter was written on 2 February 1966). After 4 months, in May, he received a reply, saying that he had to apply to the Department of External Affairs: they were the only ones who could permit correspondence with foreign embassies. Once the Department of External Affairs requested them to deal with the matter—they would do so immediately.

My heart simply aches, to see Burma in such a state. They are charming people, and there is no doubt that Burma could become an economically prosperous country. Their people are of a liberal and pleasant character.

In the evening, in the hotel's main ballroom—the annual Rotary Club reception is being held. Two hundred people in evening clothes, with music, drinks, and (Western) dancing.

3 September 1966
Rangoon

In the morning, I receive a telephone call from the French ambassador, telling me that the Chinese ambassador[24] is not prepared to meet with me, but that the North Vietnamese consul general—there is no ambassador here, only a consulate—is prepared to see me, at 4.00 pm.

I arrive at the appointed hour: he is there waiting for me, with his secretary and an interpreter. He asks me what I want, and I tell him that, on my way to Vietnam, in Paris, I had requested a visa for Hanoi but had not received a reply. I wanted to return to this request. He asks why I wanted to do this. I tell him that I want to study the Vietnam problem from a political, military, and personal point of view. I have lived with the Americans for 5 weeks and now want to be escorted to the Viet Cong army, to observe their method of fighting and understand their point of view. I said that I have three options:

1. I could be attached to Viet Cong fighting units.
2. If this is impossible, a visit to Hanoi.
3. If this too is not permitted, to speak with Viet Cong commanders—anywhere (Cambodia, Rangoon, etc.). If I could be seconded to their army—I would write about it afterwards, without revealing any military secrets.

The discussion lasts about an hour. He asks about my impressions of the war in Vietnam. I say:

1. The American army will not tire—their position is different from that of the French army more than a decade ago—this is an easy war for the Americans.
2. The Americans will win in a war with large units.
3. If the Viet Cong stop fighting in large units and go over only to guerilla warfare—they can continue this for a long time. I add that I intend to publish my impressions, without giving away technical details of the Viet Cong army. I will do the same with the American army and not publish any information I consider to be secret.

He promises to transmit my request to Hanoi with his recommendation, and to reply (if a reply will be forthcoming) via the French embassy (in Rangoon or Tel Aviv).

The North Vietnamese consul gives an impression of intelligence, with an animated gaze while he asks questions, looking me straight in the face.

At the conclusion of the discussion, I mention that there is also a technical question—he has to send a telegram to Hanoi, which would cost about $12.00, because it has to contain at least forty words.

I suggest that I pay him in dollars, local currency, travelers check, etc.—But when I take my pocketbook out, he reconsiders and says that it would be better if I pay after having received an answer: he assumes that the French ambassador would be prepared to pay for me (he regrets that a long consultation between him and his advisors takes precedence over our discussion).

While I am talking with the director of propaganda and ask him for a letter of recommendation for Namsang—he responds with pathos: 'You, General Dayan, need a letter of recommendation for Namsang? Your own experts established this settlement, etc. etc.'[25]

In an interview with a representative of Agence-France Presse, I tell him whom I met, and add that I am sorry not to have been successful in meeting people, and seeing the places that I wanted to see in Burma, because, to date, I have not been allowed to speak with, and not to travel without a permit from, the Department of External Affairs.

It has poured with rain all day. I am returning to Bangkok and, if I get a place on the plane, will return to Tel Aviv tomorrow. It is exactly 2 months since I left on 3 July1966. I have cancelled my visit to Cambodia and my trip to Washington on the advice of Disnatchik and M. Perlman[26] (today I receive a telegram from Bangkok that the American ambassador there has informed our ambassador that General Johnson is preparing a luncheon in my honor in Washington on 14 September, and, at 3.00 pm the same day, McNamara will meet with me).[27] The day before yesterday

I inform Washington, via their military attaché in Rangoon, that I have received an unexpected telegram from Israel and must return immediately, so I am cancelling my trips to Cambodia and Washington. Once in Israel, I will inform them via the American embassy if—and when—I will come to Washington.[28]

Endnotes

1 General Sankadis Tangkitcharoen. Because Thai surnames are longer and more complicated that first names, people are referred to by their first names with appropriate honorific (see notes 2, 13, 14).

2 Air Vice Marshall Air Vice Marshal Daval Preechacharoen.

3 During the Vietnam War, Laos, despite being officially neutral, became a covert battleground where the United States conducted a 'Secret War' against Communist forces, primarily by heavily bombing the country to disrupt the North Vietnamese supply lines running through the Ho Chi Minh Trail, which traversed through eastern Laos; this resulted in Laos being one of the most bombed countries per capita in history, with devastating civilian casualties and lasting environmental damage.

4 Thailand and Burma have a long history of conflict and dispute.

5 Chapter 1, note 48.

6 Zhou Enlai, Chinese statesman, diplomat, and revolutionary who served as the first Premier of the People's Republic of China from 1954 until his death in 1976.

7 Chapter 1, notes 10, 69; chapter 3 note 23.

8 Chapter 1, note 44.

9 The Mekong Committee, formed in 1957, was the precursor to the Mekong River Commission.

10 Chapter 2, note 14.

11 Chapter 1, note 47.

12 Israeli diplomats or project assistants.

13 Thais refer to white foreigners as *farangs*. The term is not derogatory.

14 General Prapat Kittipong.

15 Sarasin Vejjajiva.

16 Chapter 5, note 28.

17 Translator's comment. This section is unique. At that time Burma was practically closed to foreigners except for some members of the press, and the diplomatic corps. Dayan obtains his visa through U Thant.

18 Burma (Myanmar) is an ethnically diverse country with 135 distinct ethnic groups.

19 I could not find any of these names.

20 Term of respectful address to 'master.'

21 Brigadier General Aung Gyi (1919–2012), Burmese military officer and politician.

22 See https://www.jstor.org/stable/40394388 (accessed 24 July, 2025).

23 Mior Hon Kin, prominent figure in the Burmese government.

24 Must mean the Red Chinese Ambassador. In all other cases when Chinese officials are mentioned, Dayan is careful to add Formosa (Taiwan).

25 The Namsang project in Burma, run in the late 1950s and early 1960s to engage demobilized soldiers in establishing a series of cooperative villages, was modelled on Israeli settlements with Israeli technical and other assistance.

26 Editors of *Ma'ariv*, the newspaper that had sent him on the trip. They were obviously not prepared for him to visit Hanoi and meet with the Viet Cong.

27 Chapter 1, note 10.

28 It is difficult to interpret this as anything else but a snub at the highest levels of US government.

AFTERWORD

General Dayan never returned to Washington, D.C. for follow-up meetings but chose to return to Israel immediately, via Bangkok and Asia. Within one year after his return to Israel, he led Israeli forces fighting in the city of Jerusalem during the Six-Day War of June, 1967.

Dayan published his autobiography, *Story of My Life*, in English in 1976. He mentioned his time in Vietnam but did not provide and detailed observations that would appear in the 1977 Hebrew edition of his Vietnam diary.

Upon his death on October 23, 1979, Dayan left his literary estate to his second wife, Rachel. It is with her gracious permission that this diary has been published in English.

PHOTOGRAPH SOURCES AND PERMISSIONS

Note: The original Hebrew edition has some photographs not used in this edition. They also are from the Dan Hadani collection, National Library of Israel.

Cover:
Photographer: Israel Press and Photo Agency (I.P.P.A.) / Dan Hadani collection, National Library of Israel / CC BY 4.0. (990044326700205171).jpg
Images used with permission.

Opposite title page:
https://commons.wikimedia.org/wiki/File:Dan_Hadani_collection_(990044326750205171).jpg

Chapter 1.
Photographer: Israel Press and Photo Agency (I.P.P.A.) / Dan Hadani collection, National Library of Israel / CC BY 4.0. (990044326700205171).jpg
Image used with permission.

Chapter 2.
Alamy photograph, ID:2Y65WC1. Used with permission.
Alamy photograph, ID:2Y65WD8. Used with permission.

All other images:
Photographer: Israel Press and Photo Agency (I.P.P.A.) / Dan Hadani collection, National Library of Israel / CC BY 4.0. (990044326700205171).jpg
Images used with permission.

Images also available copyright free through Wikimedia Commons at:
https://commons.wikimedia.org/w/index.php?search=moshe+dayan+vietnam&title=Special%3AMediaSearch&type=image

ABOUT THE AUTHOR

David K. Shipler wrote of Moshe Dayan in Dayan's obituary of October 17, 1981 in *The New York Times*, 'Moshe Dayan was a general who made war, a diplomat who made peace and a political figure who stirred repeated controversy in his homeland.' Dayan was each of these things—and more.

Born on May 20, 1915 in Kibbutz Degania Alef, near the Sea of Galilee and died October 16, 1981 in Tel Aviv. Charismatic and controversial, Dayan's life was consumed by military and political service in the cause of Israeli nationalism. During the 1936–1939 popular uprising by Palestinian Arabs in Mandatory Palestine, Dayan joined the special Jewish police forces and also participated in the night squads led by British commander Orde Wingate. A member of the Haganah, Dayan was imprisoned by the British for two years in Acre. While then fighting with the British in Syria against Vichy French forces, he was struck by a sniper's bullet, resulting in the loss of his left eye and subsequent adoption of the iconic eye patch.

During the 1948 War of Independence, Dayan oversaw the defense of the Jordan Valley and was subsequently assigned as commander of the Jerusalem front. He served as the head of both the Northern and Southern commands of the Israel Defense Forces. From 1953–1958, he served as the fourth Chief of Staff of the Israel Defense Forces. From 1956–1964, he served as Minister of Agriculture.

He was appointed Minister of Defense on June 1, 1967, very shortly before the eruption of the Six-Day War. As Chief of Staff during the 1956 Sinai Campaign and as Minister of Defense during the 1967 Six-Day War, Dayan led the IDF to significant and stunning military victories resulting in popular support. But in the aftermath 1973 Yom Kippur War, he shared the blame for operational mistakes, though the official

Agranat Commission Report largely exonerated him. He was later skillfully involved with peace negotiations with Egypt. He was a major contributor to the 1978 Camp David Accords signed by Egyptian president Anwar Sadat and Israeli prime minister Menachem Begin. Dayan resigned his position as Minister of Foreign Affairs in 1980 and was elected to the 10th Knesset for the Telem party that he founded and led.

In addition to *Vietnam Diary*, he authored several other books and publications, among them: *Diary of the Sinai Campaign* (1966), *Mappah Hadasha-Yahasim Aherim* (1969), *Moshe Dayan: Story of My Life* (1976), and *Breakthrough: A Personal Account of the Egypt-Israel Peace Negotiations* (1981). A professed (and controversial) amateur archaeologist, he also published *Living with the Bible: A Warrior's Relationship with the Land of His Forebears* (1978).

ABOUT THE EDITOR AND TRANSLATOR

Peter Appelbaum is Emeritus Professor of Pathology at Pennsylvania State University. During his retirement years he has published fifteen books and translations about Jews in Central Europe during World War I and thereafter. These include five translations from the original Hebrew of works by Avigdor Hameiri including *Hell on Earth* and *The Great Madness*. He was awarded the (London) *Times Literary Supplement*—Risa Domb Porjes Prize in 2019 for his translation of Avigdor Hameiri's *Hell on Earth*. He has also published three translations of work by Theodor Lessing from German into English.